Speleology

Front cover: *Stalactites in Windeler Caverns, California.*

Back cover: *The eyeless cave crayfish* Orconectes pellucudus, *Tennessee*

Caves are fragile in many ways. Their features take hundreds of thousands of years to form. Cave animals such as blind fish are rare, and they always live in precarious ecologic balance in their underground environment. Cave features and cave life can be destroyed unknowingly by people who enter caves without informing themselves about cave conservation. Great, irreparable damage has been done by people who take stalactites and other mineral features from caves, and who disturb cave life such as bats, particularly in the winter when they are hibernating. Caves are wonderful places for scientific research and recreational adventure, but before you enter a cave, we urge you first to learn about careful caving by contacting the National Speleological Society, 2813 Cave Avenue, Huntsville, Alabama 35810 (telephone 205-852-1300; e-mail nss@caves. org).

Speleology

Caves and the Cave Environment

GEORGE W. MOORE
NICHOLAS SULLIVAN, F.S.C.

Prepared in cooperation with the National Speleological Society

Illustrated by John C. Schoenherr

Revised Third Edition

Cave Books   St. Louis

Copyright © 1964, 1978, 1997 by George W. Moore and Nicholas Sullivan, F.S.C.
All rights reserved.

Library of Congress Cataloging-in Publication Data

Moore, George William, 1928–
Speleology—Caves and the Cave Environment/by George W. Moore and Nicholas Sullivan, F.S.C.—3rd edition.
p.   cm.
Bibliography: p.
Includes index
1. Speleology. I. Sullivan, Gerardus Nicholas, 1927–   joint author. II. National Speleological Society. III. Title.
Library of Congress Catalog Card Number: 96-85614
Hardback: ISBN 0-939748-46-0
Paperback: ISBN 0-939748-45-2

This book is a new revised edition of the work first published in 1964 by D. C. Heath and Company, and in revised form in 1978 by Zephyrus Press, Inc. A Japanese language edition was published in 1973 by Tsukiji Book Company.

Manufactured in the United States of America
Cave Books, 756 Harvard Avenue, St. Louis, Missouri 63130

Contents

Preface ix
    Cave Exploration x
    Conservation of Caves and Cave Animals xi
    Acknowledgments xii
    Metric-English Equivalents xiii

1: Introduction 1
    Caves as Natural Laboratories 2
    Longest and Deepest Caves in the World 3

2: Origin of Caves 7
    Caves in Limestone 7
    Formation by Slowly Moving Water 10
    Distribution of Cave Passages 11
    Enlargement of Cave Passages 13
    Origin at the Top of the Water-Saturated Zone 14
    Stages of Limestone-Cave Evolution 18
    Vertical Shafts 19
    Caves Formed by Hydrogen Sulfide Gas 20
    Dating Caves and Cave Deposits 22
    Paleomagnetism of Cave Silt 23
    Karst 28
    Caves not in Limestone 30

3: Characteristics of the Underground Atmosphere 33
    Controls of Cave Temperature 33
    Caves Containing Perpetual Ice 36
    Relative Humidity 37
    Air Currents Caused by Barometric Change Near Entrances 38

Chimney and Reverse-Chimney Effects 38
Breathing Caves 39
Ebb and Flow Springs 41

4: Growth of Stalactites and Other Speleothems 47

Importance of Soil Carbon Dioxide 48
Stalactites and Related Deposits 49
Stalagmites and Related Deposits 55
Deposits Formed by Seeping Water 59
Deposits Formed in Standing Water 66
Ancient Climate Recorded by Speleothems 68
Cave Minerals 70

5: Behavior and Products of Cave Microorganisms 79

Characteristics of the Microflora 80
Saltpeter 82
Manganese and Iron Minerals 83
Moonmilk 85
Medical Use of Cave Actinomycetes 89
Harmful Microorganisms 90
Relation of Microorganisms to Cave Food Chains 91

6: Habits of Cave Animals 93

Degree of Adaptation to Caves 94
Troglobites 95
Vertebrate Cave Animals 105
Segregation of Animals 114
Cyclic Behavior Patterns 116
Cave Food Web 118
Cave Ecology Based on Hydrogen Sulfide 119
Troglobites From the Sea 121

7: Evolution of Blind Cave Animals 125

Colonization of Caves 125
Intergradation with Surface Animals 128
Reproductive Separation of Cave and Surface Relatives 129
Regressive Characteristics 130
Mutations Unchecked by Natural Selection 131

8: Uses of Caves 135
    Human Ancestors 135
    Cave Art 137
    Early Americans 140
    Modern Uses of Caves 145

References and Related Reading 149

Caves in the United States Open to the Public 163

The Authors 170

Index 171

Preface

SPELEOLOGY treats both the physical and biological aspects of the cave environment. It is one of the newest of the sciences, and has undergone vigorous expansion in the past few decades bringing to light many new questions as well as new facts. The new facts make the understanding of some cave processes easier now than ever before.

Speleology consists mainly of basic research and embraces several branches of biology and geology, as well as chemistry, meteorology, and soil science. Because it cuts across so many discrete fields of science, it can avoid the pigeonholing that sometimes hampers innovation in other fields.

Since publication of the second edition of this book in 1978, research progress has been particularly rapid. In this third edition, we have added new results and corrections at many places to topics that we dealt with previously. Other material new to this edition includes the origin of some large caves, such as Lechiguilla Cave, New Mexico, from dissolution by sulfuric acid derived from oil pools; establishment of a cave food web as at Movile Cave, Romania, based on hydrogen sulfide that requires no energy input from the surface; building of a complete 2-million-year time scale for Mammoth Cave, Kentucky, based on repeated reversals of the Earth's magnetic field recorded in cave silt; dating of cave silt by fossil teeth from rapidly evolving meadow mice; new concepts of the evolution of blind and nonpigmented cave animals derived from evidence in tropical caves; and evidence that dangerous sinkhole collapse can be caused by pumping down the regional water table. Also, the list of caves in the United States open to the public has been revised and brought up to date.

Our chief objective in this book is to analyze the processes operating in the cave environment, that is, to develop the underlying principles of speleology. We have also included enough descriptive material to illustrate and to support these principles. Many problems remain to be solved, and some of our tentative conclusions are not endorsed by all speleologists. We do not expect our views to be accepted without question; our hope is rather that they will stir controversy and point the way to new research.

Cave Exploration

IN THE FOLLOWING CHAPTERS, we deal only with the results of cave investigation, but since this book will be an introduction to speleology for many readers, we would like to point out some of the hazards of cave exploring.

Despite a widely held belief, people rarely come to great harm simply by getting lost in a cave. However, the notion that one can keep from getting lost by unwinding a ball of string from the entrance as Tom Sawyer and Becky Thatcher did in Mark Twain's famous book, is nonsense. Most balls of string are only a few hundred meters long and therefore would barely reach beyond the twilight zone. A competent speleologist carries a headlamp mounted on a hard hat, and two additional sources of light, such as a flashlight and a tiny 5-day light-emitting diode, in case the first light fails. In fact, a careful observer with a good memory and a good light is less likely to get lost in a cave than in the woods. Distinctive markings on the walls, floor, and ceiling, provide reference points that explorers can recall when retracing their steps.

The greatest hazard in cave exploration is falling. Cave explorers rarely stumble into holes. Most falls occur when inexperienced explorers use old ropes or ladders that they have found in caves, or when they descend ropes improperly. Ropes and wooden ladders left by previous visitors should never be used, because decay organisms that abound in caves rapidly weaken them without greatly changing their outward appearance.

When descending a rope, one should use a special braking technique, known as *rappelling*, to avoid sliding too fast. This technique, and another useful one known as *belaying*, in which a climber is secured to a safety rope held by a companion, who can thereby catch the climber if he falls, can be learned from local members of a rock-climbing club or the National Speleological Society. But these techniques give only a false sense of security if they are not skillfully used. They should not be used without previous instruction.

Conservation of Caves and Cave Animals

Caves have great beauty and scientific value in their natural state, but unfortunately they are very easily damaged. Damage to caves is all the more unfortunate because they constitute a natural resource that is—like redwood forests, for example—comparatively rare. Even the largest caves are small compared with major scenic features on the surface of the Earth. The total volume of all the limestone caves in the world is probably less than 50 cubic kilometers—less than one 10-millionth of the total volume of the ocean.

A dramatic interplay takes place among the various forms of life that inhabit caves. These forms must be able to share their small world with members of their own species and of other species, and they must also be able to tolerate the physical and chemical influences of their surroundings. The animal population within most caves is extremely small, and many cave species are on the verge of extinction. The removal of just a few individuals can seriously disturb the balance of nature and contribute to the extinction of a species.

In caves that contain hibernating bats in winter, one should refrain from entering certain passages, because even the movement of people can create enough of a disturbance to disrupt the sleep of the bats. There is evidence of high mortality in bat caves that people visit frequently.

Cave explorers should avoid leaving any waste in a cave. It is especially desirable to remove spent carbide, because it disrupts the delicate chemical balance that is so important to cave life. Plastic bags are useful for carrying out carbide and other waste.

Explorers should make no markings in a cave except those needed for surveys, and these should be easy to remove. Guide arrows should not be marked on the walls. Not only do they deface the cave, but also they can create a hazard by confusing later explorers.

The mineral deposits in caves are quite fragile. When a stalactite, for example, is broken, the break will never heal. A stalactite once broken is marred forever, and no one can ever again see it in its natural state.

Because all natural features in caves are important in maintaining the cave environment, we advise against collecting for display purposes. Even taking out previously broken mineral specimens or dead animals may encourage others to collect live animals or to break off stalactites from the walls. Specialists can justifiably collect for research, but they should, and usually do, hold their collecting to a minimum.

In few places on this planet is an environment available for study that is so uncomplicated and so free from contamination as that of caves. Let us do all we can to preserve this unique but fragile environment.

Acknowledgments

THIS BOOK was written in cooperation with the National Speleological Society. Members of the Society assisted in many ways—by providing data, by checking parts of the manuscript for accuracy, and by making available their unpublished ideas. Technical editors of one or more chapters of the various editions were Thomas C. Barr, Jr., Frank C. Calkins, Robert F. Heizer, Peter Kurz, Warren C. Lewis, Ellen J. Moore, Thomas L. Poulson, Victor A. Schmidt, Patty Jo Watson, and William B. White. Richard A. Watson reviewed the entire manuscript of this edition.

GEORGE W. MOORE
NICHOLAS SULLIVAN, F.S.C.

Metric-English Equivalents

INTERNATIONAL UNIT		ENGLISH EQUIVALENT
centimeter (cm)	=	0.394 inch
centimeter per second (cm/s)	=	0.00328 foot per second
cubic centimeter (cm^3)	=	0.0612 cubic inch
cubic kilometer (km^3)	=	0.240 cubic mile
cubic meter (m^3)	=	35.3 cubic feet
degrees Celsius (°C)	=	[(1.8 × °C) +32] degrees F
gram (g)	=	0.0353 avoirdupois ounce
kilogram (kg)	=	2.20 pounds
kilometer (km)	=	0.621 statute mile
liter (L)	=	0.264 U.S. gallon
liter per second (L/s)	=	0.0353 cubic foot per second
meter (m)	=	3.28 feet
metric ton (t)	=	1.10 short ton
micrometer (μm)	=	0.0000394 inch
millimeter (mm)	=	0.0394 inch
pascal (Pa)	=	0.00000987 atmosphere
square kilometer (km^2)	=	0.386 square mile
square meter (m^2)	=	10.8 square feet

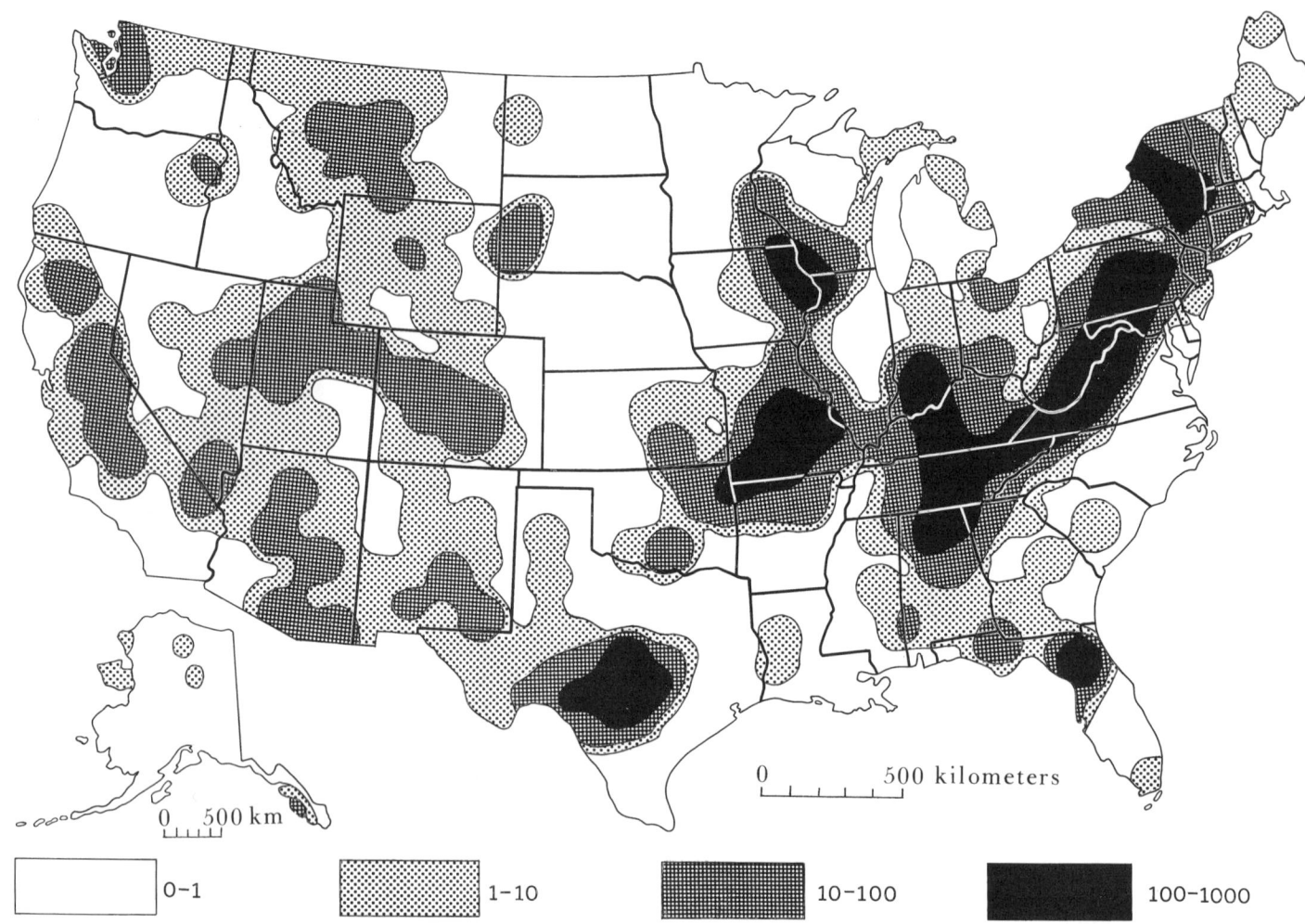

Known limestone caves of the United States. The patterns indicate the number of caves per 25,000 square kilometers. The edge of the lightest pattern lies 90 kilometers (the radius of a 25,000 square kilometer circle) beyond the nearest cave. Several dissolution caves in former reefs are known in Hawaii.

1: Introduction

PEOPLE have long used caves as dwelling places, as burial sites, as storehouses, and as places of worship. Many societies attach an almost mystical significance to caves. Prehistoric cave drawings have been discovered in various parts of Europe and Asia that are as old as 30,000 years. These drawings, many of them extremely well preserved by the cave environment, display great technical skill and artistic talent. Clay statues and various other types of sculpture have also been discovered in these primitive art galleries. Some of the works of the long-dead artists who labored deep inside caves may have been magic for hunting or for protection against enemies. Perhaps we are not so far removed from these ancestors as we like to think, for within the past few decades some people have looked again toward caves, this time for possible protection from nuclear weapons.

Only recently have we come to realize that caves have an interest apart from the light they throw on the lives of our remote ancestors, or of the animals whose fossil remains are preserved in them. Modern exploration of caves began several hundred years ago, particularly in Europe, but not until the middle of the 19th century did caves become the objects of intensive scientific investigation. Most of the pioneers in the study of caves, who came to be known as *speleologists* (from the Greek words *spelaion,* meaning cave, and *logos,* meaning study) were European geologists and biologists. These scientists soon realized that the vast underground areas offered extensive mapping problems, that no one understood the origin of caves, and that the strange blind animals that inhabited them were virtually unknown.

Today the study of caves has spread throughout the world. Although the chief efforts in the past were directed toward caves in Europe and North America, work is now well advanced on caves of the other continents.

Speleology is no longer a highly specialized pastime in which we are incidentally studying unusual but relatively unimportant facets of nature. As caves have become better known, we have realized that they can broaden our understanding of the interaction of certain biologic and geologic processes that have been shaping our planet and its inhabitants for hundreds of millions of years. Thus, the study of caves is an important means of understanding our world.

Caves as Natural Laboratories

FOR THE GEOLOGIST interested in the structure and processes of the Earth, caves hold keys to the solutions of many problems. The formation of certain mineral deposits, the processes of soil formation, and the rate of water movement through rocks, are just a few of the subjects on which the study of caves can throw new light.

In parts of southern France, communities have been able to tap previously unknown water supplies after cave investigations revealed large quantities of pure water underground. In the United States, cave studies have helped to show hydrologists how underground water supplies can become contaminated. In some instances, for example, scientists found that sewage dumped into one stream returned to the surface again some distance away in another stream that was being used as a source of drinking water. Disease had consequently been spreading in a way that suggested black magic, but the mystery was cleared up when these underground connections between watercourses were discovered.

Recent developments in the physiochemical study of cave mineral deposits have established an important new line of evidence about the past climates of continental areas. Samples of dripstone can be dated by measuring radioactive decay, and their temperature of formation can be estimated by isotopic analysis. A cave's temperature approximates the average annual temperature at the surface, so cores drilled into dripstone deposits can provide an accurately dated sequential record of past temperature changes that can be extrapolated back for several hundred thousand years. This supplements information from surface weather records, which cover too short a time span to define long-period cycles of weather change.

For the biologist, both the fauna and the flora of caves present many intriguing problems. How do cave-dwelling organisms adapt to a realm of total darkness? How does animal life manage to subsist in spite of an extremely limited food supply? How do cave food chains differ from those of the surface as a consequence of the lack of sunlight so important to the growth of chlorophyll-bearing plants? In studying these

Longest and Deepest Caves in the World
(Compiled by Robert Gulden, June 1996.)

CAVE	COUNTRY	LENGTH (Kilometers)	DEPTH (Meters)
Mammoth (Ky)	United States	563	116
Optimisticeskheskaya	Ukraine	192	20
Jewel (S Dak)	United States	170	179
Hölloch	Switzerland	166	872
Lechuguilla (N Mex)	United States	142	490
Siebenhengste Hohgant	Switzerland	135	1324
Wind (S Dak)	United States	126	198
Fisher Ridge (Ky)	United States	126	88
Ozernaya	Ukraine	111	—
Air Jernih	Malaysia	102	355
Ojo Guarena	Spain	97	—
Coumo d'Hyouernedo	France	95	1018
Zolushika	Moldova	86	30
Purificación	Mexico	85	956
Hirlatz	Austria	79	1041
Easegill	United Kingdom	71	211
Raucherkar	Austria	70	725
Friars Hole (WVa)	United States	70	188
Paolo Roversi	Italy	40	1249
Madejuno	Spain	29	1255
Rebecos	Spain	22	1255
Cosanostraloch	Austria	30	1265
Berger	France	26	1271
Pierre Saint Martin	France	54	1342
Ceki	Slovenia	—	1370
Sniezhnaja-Mezhonnogo	Georgia	19	1370
Cheve	Mexico	24	1386
Lukina	Croatia	—	1393
Laminako Aterneko Leizea	Spain	15	1408
Boj-Bulok	Uzbekistan	5	1415
Trave	Spain	9	1441
Huautla	Mexico	57	1475
Vjacheslava Pantjukhina	Georgia	—	1508
Mirolda/Lucien Bouclier	France	9	1520
Lamprechtsofen	Austria	38	1537
Jean Bernard	France	20	1602

problems, biologists find that organisms living underground have rates of growth different from those of similar organisms on the surface, and that cave animals have external features different from those of their surface relatives. For these reasons, caves provide an important laboratory for the study of evolution. Cave studies have thrown doubt, for example, on the view that the blindness characteristic of many cave animals developed exclusively through the process of natural selection advanced by Charles Darwin in *On the Origin of Species*.

Investigators have found thousands of species of animals, known as troglobites, that live only in caves. Much remains to be learned about these animals' life cycles, their habits, and their life spans. Cave biologists are also interested in bats and in some larger animals that use caves merely as part-time dwelling places or as emergency shelters. The droppings of these visitors contribute substantially to the food supply of the permanent cave dwellers.

Some of the unique forms of underground life produce special chemical compounds that may have practical applications, but which are only now being carefully investigated. Until recently, even the existence of some of these organisms was unknown. For example certain of the actinomycetes—moldlike bacteria—may prove to be a potential source of new kinds of antibiotics. The recent discovery that sulfur bacteria produce vitamins of the B group has opened another important area of research. Possibly the abundant sulfur bacteria and other bacteria that occur in caves will supply useful products.

The research methods of interdisciplinary sciences such as speleology have been widely applied during the past 30 years, while a general awareness of the importance of ecologic interaction has increased. Hence, concepts developed in speleology have provided useful guidelines as the need for environmental protection has become more obvious to everyone. For example, the concept of monitoring changes in an uncomplicated microenvironment such as a cave may have great utility in sounding a warning against potentially harmful broader changes at the surface that might be masked by other changes people have caused.

In this age of interest in space exploration, it may seen paradoxical to say that the study of underground areas can contribute to space biology. But consider the fact that on a long space voyage, it will be necessary to recycle the human waste products as a source of new nutrient material. Study of the food cycles in caves can give clues to possible ways of treating these waste products.

People traveling through space live in a foreign environment. We are accustomed to the daily rotation of the Earth, with its associated 24-hour cycles of change in atmospheric pressure, in amount of light, and so forth. People have adapted themselves to these variations in their environment, but when traveling in space, and particularly on trips lasting months or years, they will be living in an environment that does not undergo these daily changes. How the absence of the cycles affects the physiology or psychology of space explorers is not yet fully known, but anyone who has traveled a long distance by air is familiar with the discomfort of jet lag. Scientists have discovered that blind cave crickets and millipedes exhibit a twice-daily tidal cycle in their activity, despite these species having dwelt in the dark for thousands of generations, and having, in fact, lost their sight and pigmentation as a consequence. Whether human life processes are likewise keyed to some effect related to tidal-gravity change, and whether the absence of this effect would be harmful on very long space voyages, can be determined only by experience. But not all of the necessary experiments must await the arrival of extended space travel. Some of them can be performed inexpensively and in relative safety in caves.

We see, therefore, that caves can serve as natural laboratories in many ways. Their study not only gives us a deeper insight into nature, but also can possibly improve our well-being, help us protect our world, and perhaps even assist us when we reach out to other worlds.

Swiftly flowing streams have cut small hollows, called scallops, in the walls of some caves. The scallops in this illustration indicate former water flow to the right, for the steep slope of every scallop is on the upstream side.

2: Origin of Caves

Caves in Limestone

LIMESTONE CAVES are many times more numerous than those in other kinds of rock. They are also unrivaled in size and especially in the distances they extend underground. Some individual rooms in limestone caves are larger than any man-made room underground. Mammoth Cave in Kentucky, for example is part of a winding series of connected passages with a total surveyed length of 563 kilometers. Carlsbad Cavern in New Mexico has one chamber whose area is 38,000 square meters—about 14 times that of a football field—and part of whose ceiling is 75 meters above the floor—about one and one-half times the height of Niagara Falls.

Chambers of such enormous volume as these are of course uncommon, and many interesting limestone caves are comparatively small. Nevertheless, these examples show what natural forces can effect in the way of underground excavation when given sufficient time.

When engineers drive a tunnel or open a mine, they work as rapidly as they can, amid a deafening clatter of machinery punctuated by the detonation of powerful explosives. Nature, by contrast, hollows out limestone caves in almost complete silence, usually with no tool other than slowly moving, weakly acidified water. However, nature has, so to speak, all the time in the world. Every large cave is probably at least a million years old. Human engineers do not have such periods of time to work with. Great lengths of time are necessary for the formation of caves, however, because natural processes are so slow.

Limestone and marble—which is limestone that has been recrystallized by heat and pressure—are composed of the mineral calcite ($CaCO_3$). The limestone and marble that now contain caves were formed in ancient seas millions of years ago by marine animals and plants that extracted calcium carbonate from sea water. Grains composed of fragments of the skeletons of these organisms, together with extremely small grains produced by microorganisms, were later compacted under pressure and cemented into firm rock. Finally, mountain-building forces uplifted these sedimentary rocks and exposed them to the air and to the dissolving power of fresh underground water.

Caves have been formed in limestone beds of widely differing ages. Researchers have long determined the relative ages of limestone formations and other sedimentary rocks by studying the fossils that these rocks contain. It is now also possible to estimate the *absolute* ages of many rocks within a few thousands or millions of years. We estimate the ages of fossiliferous rocks that contain uranium ores from the amount of lead that has accumulated in them through radioactive disintegration of the uranium. We then infer that a rock containing the same fossils, but no uranium ore, is the same age. In this way we have learned that the limestone containing Mammoth Cave, for example, is about 325,000,000 years old, and that the limestone containing Carlsbad Cavern is about 250,000,000 years old.

Age of the Limestone Containing Some Well-Known Caves
(The caves were formed less than 10 million years ago during the relatively recent Tertiary and Quaternary Periods.)

Cave	Geologic period	Approximate age (millions of years)
Florida Caverns, Florida	Tertiary	50
Caverns of Sonora, Texas	Cretaceous	100
Aven Armand, France	Jurassic	150
Oregon Caves, Oregon	Triassic	200
Carlsbad Cavern, New Mexico	Permian	250
Spanish Cave, Colorado	Pennsylvanian	300
Mammoth Cave, Kentucky	Mississippian	375
Howe Caverns, New York	Devonian	375
Perry Cave, Ohio	Silurian	425
Luray Caverns, Virginia	Ordovician	450
Lehman Caves, Nevada	Cambrian	550
Eldons Cave, Massachusetts	Precambrian	1000

The age of a cave, however, bears little relation to the age of the rocks that enclose it. Most caves are very much younger than their enclosing rock. In fact, all the important limestone caves in the world, including some in rocks laid down hundreds of millions of years ago, are less than 10 million years old.

The limestone removed in producing a cave is not simply dissolved in water. In fact, limestone is only slightly more soluble in pure water than quartz, the chief mineral of sandstone, and is less soluble than most of the minerals in such rocks as granite and basalt, which rarely contain caves formed by dissolution. Limestone caves are formed when acids attack calcite. Even such very dilute acids as those in ground water can produce caves if given enough time. We can easily demonstrate the reaction between an acid and calcite in the laboratory by dropping a chip of limestone or marble into a beaker containing dilute hydrochloric acid. The chip will dance and bubble vigorously, and if sufficient acid is present it will finally disappear.

The acid chiefly responsible for the natural dissolution of limestone to form caves is carbonic acid (H_2CO_3), produced when carbon dioxide, the universal product of plant decay and animal respiration, combines with water. Carbonic acid is weak, even at maximum concentration. The outside atmosphere contains 0.03 percent carbon dioxide, but the carbonic acid produced from this is too dilute to be very effective in forming caves. Most of the carbon dioxide responsible for creating the acid that dissolves limestone to produce caves comes from the surface soil, where large amounts are produced by decay of humus.

Carbon dioxide and water work together to remove limestone by means of this double reaction:

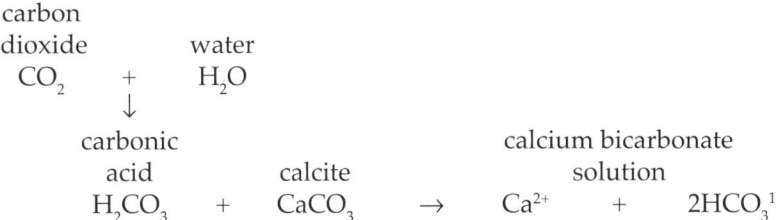

The carbon dioxide combines with water to produce carbonic acid, which in turn attacks the calcite and divides it into soluble ions. A cubic meter of water exposed to air containing 10 percent carbon dioxide, if kept in contact with limestone until the reaction ceases, can dissolve about 250 grams of calcite.

2 / Origin of Caves

In many caves, the passages follow joints that intersect nearly at right angles in the limestone. Hamilton Cave, West Virginia, a plan of which is shown here, is a good example. (*After* William Davies.)

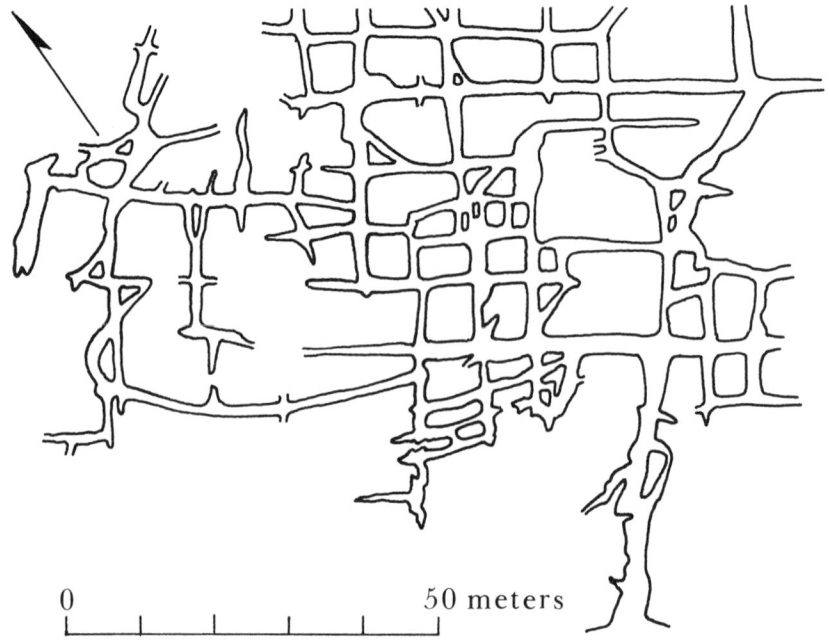

Formation by Slowly Moving Water

UNTIL the present century, scientists generally supposed that caves had been eroded by underground streams, just as valleys are eroded by surface streams. Compelling arguments against this idea, however, were offered by the Austrian geologist Alfred Grund, and by the American geologist William Morris Davis. Grund and Davis point out that the shapes of most cave passages do not resemble passages formed by downcutting streams. Cave passages usually form a network, so that cave maps often look like maps of cities with many intersecting streets. Such a grid is quite different from that eroded by surface streams, where the pattern formed by tributaries joining the main stream is often like that of a branching tree.

Scientists now believe that most caves were formed by the activity of slowly moving water in the zone below the water table, which is the level below which the rocks are water-saturated.

A second line of evidence against the hypothesis that caves are formed by the activity of underground rivers is that cave walls are generally smooth or gently undulating. The beds of fast-moving streams on limestone are never smooth. Where surface streams flow on lime-

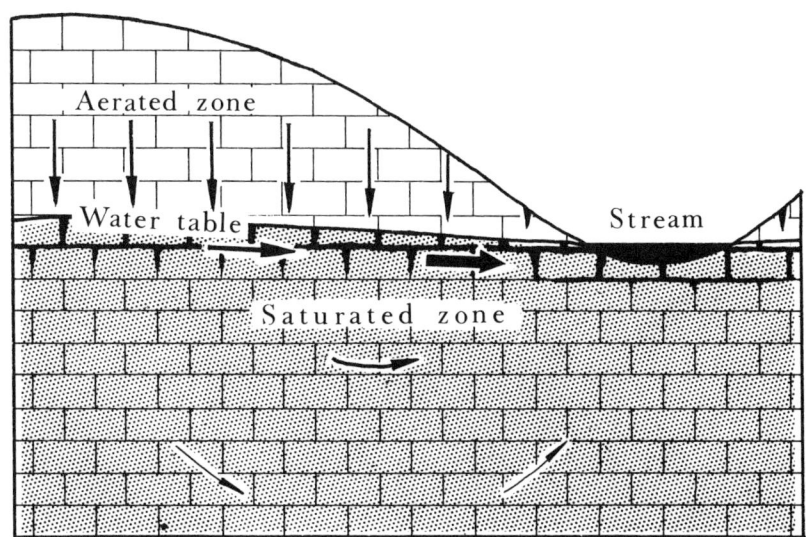

Flow of subsurface water in limestone. The water moves extremely slowly, except in a thin horizontal layer directly below the water table.

stone, or where a stream has entered a previously existing cave, the stream bed is pitted with small dissolution pockets known as *scallops*. These distinctive indentations, which are usually a few centimeters to 1 meter across, have steep slopes on the upstream side and gentle ones on the downstream side. Scallops, therefore, are sometimes useful for determining the direction of flow of former cave streams, although they are usually present only in small parts of a cave system, or not at all. Most scallops were formed by a stream that penetrated the cave late in its history, so they usually lie near the floor, below a definite high-water line. The absence of small scallops in the greater part of most limestone caves supports the hypothesis that the caves were not formed by underground streams, but by slowly moving water, and thus, by implication, below the water table.

Distribution of Cave Passages

WHERE A LAYER of limestone is interbedded with insoluble rock, cave passages appear only in the limestone. If the limestone bed is thin, it is possible to predict the directions in which undiscovered passages might be found. Besides the distribution of soluble limestone, two other factors control the distribution of cave passages—vertical to horizontal fractures in the limestone, and the water table, which forms a horizon-

Relation between the size of scallops etched into cave walls and the velocity of the water that produces them at two different water temperatures. The lines refer to the velocity one scallop length out from the wall and show that the faster the flow, the smaller the scallops.

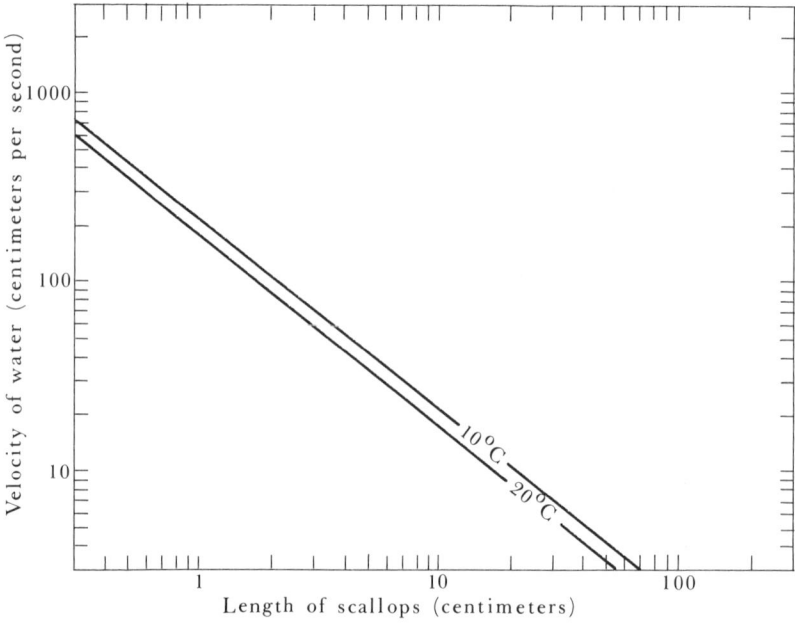

tal plane that determines the level at which many cave passages are formed.

The fractures occurring in limestone are of three types: (1) *partings*, which are parallel with the bedding; (2) *joints*, which cut across the bedding but along which there has been no displacement; and (3) *faults*, which cut across the bedding and along which it has been displaced.

Partings follow thin silt or clay layers laid down with the limestone. Joints occur in both folded and nonfolded rocks. They are thought to be caused by earth tides, which produce a gentle flexing of the rocks. Earth tides have the same cause as ocean tides—the solid part of the Earth, like the water of the oceans, is attracted by the Moon and the Sun. Earth tides average only about 30 centimeters in height. The resulting joints may be compared with fatigue cracks that form in metal when it is repeatedly bent back and forth. Joints may take thousands of years to form, because the tidal flexing occurs only twice daily, and the displacement on each joint is very small. Faults are caused by mountain-building forces that fold the rocks until they break.

Of the three types of fractures, faults are the least important to speleology, because in most cave regions they are rather rare and widely spaced. The shapes of cave passages are controlled chiefly by joints and partings that occur at least every meter in limestone beds. When the limestone beds are horizontal, cave maps often show passages following two sets of joints intersecting approximately at right angles to each other. When the limestone beds have been tilted so that they dip steeply, the main cave passages are generally elongated along the strike—the direction of the line along which the limestone beds intersect the horizontal—and short passages extend along joints at right angles to these main passages.

Enlargement of Cave Passages

THE RATE OF HORIZONTAL MOVEMENT of water in the small fractures below the water table is commonly less than 10 meters per year, whereas in cave-sized conduits it averages 0.2 kilometers per hour. The water first moves downward through cracks to the water table, then collects in joints and partings below the water table, and finally moves through the limestone to points of discharge at springs. When the fractures are small, the water becomes saturated with calcite soon after it descends below the water table, so the early stages of dissolution are very slow. During this period, flexing by earth-tidal action along the joints may help to pump the water along. Acids produced within the rocks by oxidation of sulfide minerals in the limestone may accelerate the process of dissolution in places and permit it to operate at considerable depth below the water table.

The rate of flow is at first about equal through all the joints, but as some channels grow larger than others, they take in more water and therefore grow faster. When a channel reaches a certain critical size—about 5 millimeters in diameter—it subsequently grows at such a fast rate that it greatly outpaces adjacent channels, takes nearly all the water flow, and hence grows even faster.

We are not certain why a diameter of 5 millimeters is critical in the development of cave passages. One possibility is that at this diameter the flow of water in the channel becomes turbulent. When water is sufficiently undersaturated, the start of turbulence in a limestone channel greatly increases the effectiveness of the dissolution process. The action is similar to stirring sugar in a liquid to dissolve it—fresh undersaturated solution is constantly being brought into contact with the solid.

William White has noted that among waters undersaturated with respect to calcite, a great acceleration in rate of dissolution occurs at a critical level of saturation. Water slightly undersaturated dissolves calcite very slowly, whereas water only a little more undersaturated dissolves it very rapidly. White's studies suggest that the critical level of undersaturation usually occurs at a channel diameter of approximately 5 millimeters—the same diameter at which turbulence also begins. When critical undersaturation and the start of turbulence affect a channel, the channel robs nearby fractures of their flow. The channel, then being enlarged at an accelerating rate, ultimately grows into a cave passage, while its remaining neighbors never exceed 5 millimeters in diameter.

Origin at the Top of the Water-Saturated Zone

ALTHOUGH THE CHARACTERISTIC NETWORK PATTERN and the absence of scallops on the walls show that most caves were formed by slowly moving water below the water table, the exact depths at which caves form are not shown by these types of evidence. However, another observation—that many cave passages are horizontal—suggests that the horizontal water table may have an influence on the origin of caves.

In caves where the limestone beds themselves are horizontal, this might seem to be a sufficient reason for the horizontality of the cave passages, but cave passages are horizontal even in areas where the limestone beds are folded or steeply dipping. Moreover, recent work in very large cave systems in nearly flat-lying limestone, where speleologists formerly thought that the cave passages merely followed the most soluble beds, has shown that even here the passages cut slightly across the bedding of the limestone. These passages are controlled by a nearly horizontal plane that is not precisely parallel to the bedding. In view

Many caves are nearly horizontal, as shown in this profile of Lehman Caves, Nevada. Such caves are thought to have formed directly below a horizontal water table.

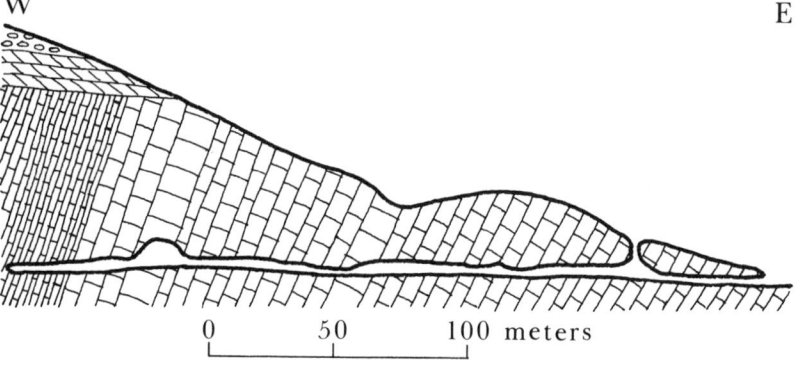

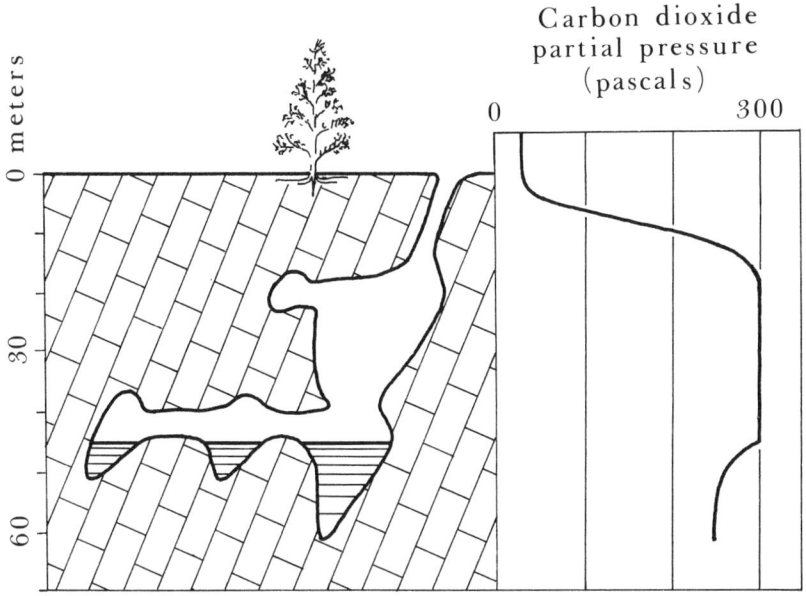

Carbon dioxide content of the water and air in Black Chasm Cave, California, and of the air above the cave, in May 1962. The underwater samples were collected by a diver. The carbon dioxide of the water, and therefore its carbonic acid content, is greatest near the water surface.

of this evidence, many modern investigators agree that most limestone caves have been formed in a relatively thin horizontal zone directly below the water table.

We believe that cave passages form directly below the water table because of a combination of several factors: (1) the carbon dioxide content in this zone is relatively high; (2) unlike the downward-moving water above, the water below the water table is in contact with the limestone long enough to become fully saturated with calcium carbonate; and (3) because of a nonlinear relation between carbon dioxide content and calcite solubility, mixed downward-percolating water and ground water has a greater dissolution capacity than either of these waters alone.

To clarify these factors further: Water containing carbon dioxide derived from the soil moves downward so rapidly in the aerated zone that it may dissolve only a small fraction of the limestone that it could dissolve if given enough time. When this water discharges into the main body of ground water below the water table, it still contains unneutralized carbonic acid. But even if the water from the aerated zone is saturated with calcite, and the water below the water table is too, their mixture has the capacity to dissolve more limestone. That is, the mixing of two waters that have different carbon dioxide contents, both saturated with calcite, leads to an undersaturated mixture that

Typical changes in the chemical characteristics of rainwater as it moves through the soil, through the limestone, into a cave, and finally back to the surface from a limestone spring.

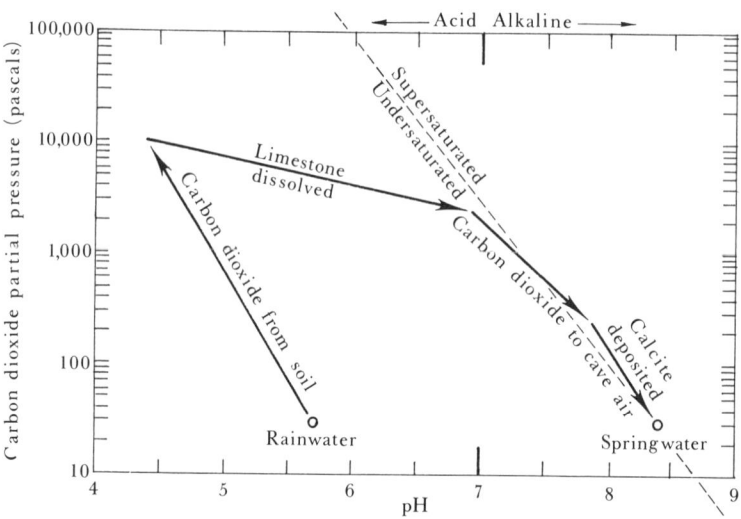

has excess carbon dioxide and thus is capable of dissolving more limestone. This phenomenon results from the fact that calcite dissolution is more sensitive to an increase in carbon dioxide content at low carbon dioxide levels than at high levels. Therefore, any mixing enhances dissolution.

As a result of the excess carbon dioxide in water near the water table, the small cavities directly above the water table acquire a high and uniform carbon dioxide content. This carbon dioxide diffuses downward from these cavities into the upper layer of the ground water, imparting an almost constant dissolving capacity to the water near the top of the water-saturated zone—a capacity much greater than that of the water far below the water table.

Master channels therefore tend to form in a limited vertical zone directly below the water table. As these channels become larger, nearly all the water movement below the water table is in this zone. The dissolved calcium carbonate is carried along a very gentle slope at the top of the saturated zone, sometimes for a distance of many kilometers, to outlets along stream valleys.

Water that enters a sinking stream at the surface usually has a low content of dissolved limestone and a low concentration of carbon dioxide (equal to that in the surface atmosphere). Such water cannot dissolve much more limestone. When it enters a water-filled conduit that is insulated from the outside atmosphere, however, it absorbs more carbon dioxide from the main ground-water body, and the amount of limestone that it can take into solution is greatly increased. This helps the throughgoing conduits to grow much faster than the intervening narrow dissolution cavities.

This cave-forming process may continue for thousands of years. Only two things can stop it—a lowering of the water table, or the exposure of the cave system to the open air by surface erosion. The lowering of the water table drains the cave, which means that dissolution either must cease or must continue at a lower level, perhaps until a lower system of cave passages has formed. The opening of a cave entrance or other air passages usually marks the end of the cave-forming process. Because the entrance allows ventilation to begin, the high partial pressure of carbon dioxide can no longer be maintained in the cavities above the water table. The excess carbon dioxide is dissipated, the water quickly becomes saturated or even supersaturated with calcite, and the dissolution process ceases. Usually, in fact, this change marks the beginning of a reverse process, the deposition of calcite in the form of stalactites and other speleothems. At approximately the moment when a cave becomes accessible to people, it begins to be decorated by the features that mean most to the average visitor. But that moment is

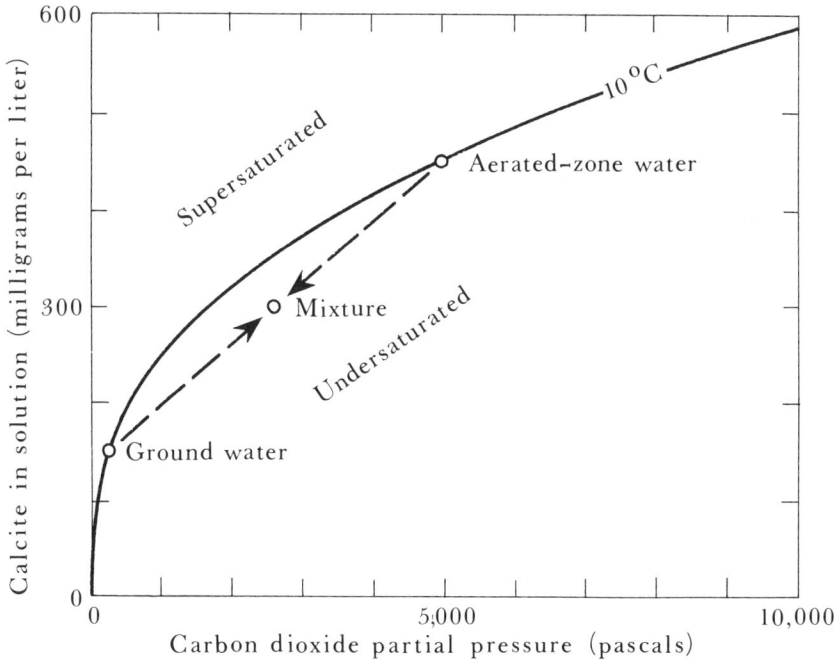

The mixing of downward-percolating water with ground water at the water table invariably results in a solution undersaturated with respect to calcite, even if both waters were initially saturated. This mixture dissolution is believed to be one of the principal reasons why cave formation takes place directly below the water table in the layer where such mixing usually occurs.

only the beginning of the process; the elaborate decorations in our most beautiful caves have been forming for thousands of years and will continue to form as long as water continues to percolate downward from the surface.

Stages of Limestone-Cave Evolution

THE FORMATION OF LIMESTONE CAVES is directly related to the development of the overlying land surface. Many examples of nearly horizontal cave passages occur in areas of modern mountain-building activity, where earthquakes are common and young sediments have been noticeably tilted. For example, the caves at Lehman Caves National Monument have nearly level passages, even though the mountain range in which they lie has been conspicuously tilted within the past 5 million years. This shows that limestone caves are, geologically speaking, very young and short-lived. Never more than a few million years intervene between the initial stages of a cave and the time when it is destroyed by the collapse of its roof.

Arthur Palmer emphasizes that cave dissolution intergrades continuously between passages at the water table and at active streams that lead downward from sinkholes. In mature caves, however, relict intermediate passages at former water-table levels can confuse the intergradation. The streams from sinkholes wander through the relict passages and create a branching pattern of downcut canyons that lead to trunk passages. During heavy rains, water containing only a small amount of dissolved limestone floods along the sinkhole streams and backs up from river level into the water-table passages. Carbon dioxide from percolation charges the water in both places, and during such floods it removes limestone at maximum rate.

Limestone caves usually develop in four stages: (1) the initial enlargement of joints and partings by ground water in the water-saturated zone; (2) the development of master channels directly below the water table during a period when the altitude of the water table is relatively stable and a high partial pressure of carbon dioxide exists at the top of the water-saturated zone; (3) a transitional stage during which nearby streams have cut down to the point where their seasonal fluctuations strongly affect the level of the water table in the cave, sometimes introducing river silt into the cave system; (4) the further lowering of the water table and downcutting of the surface until an opening to the surface is created, the carbonic acid content of the cave water

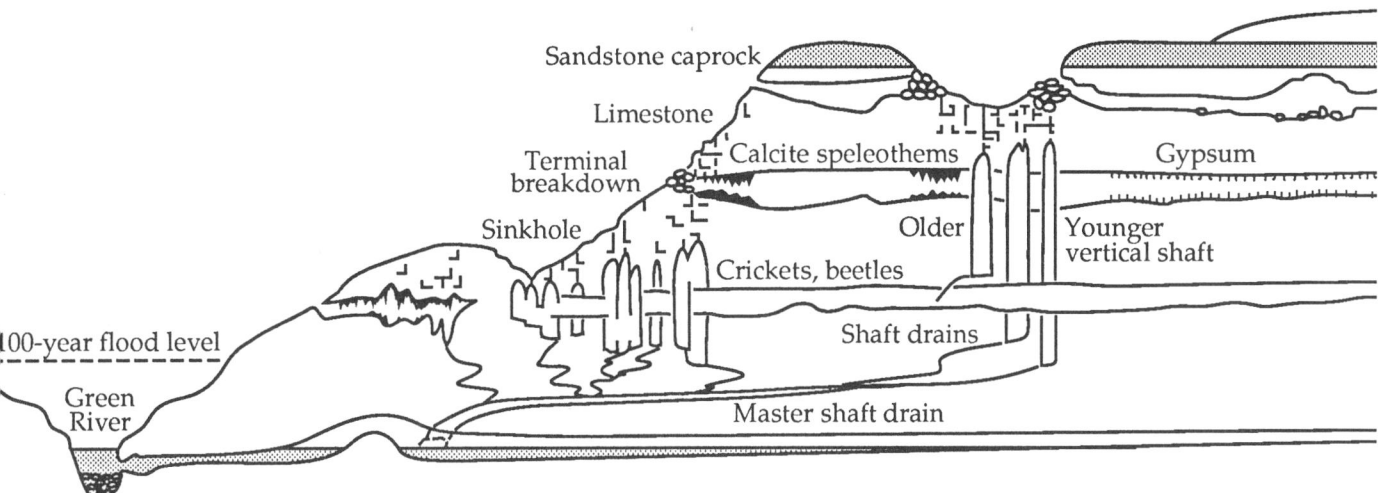

consequently becoming so low that the water ceases to dissolve the limestone. Meanwhile, the acid content in the aerated zone between the soil and the cave remains high. In the final phase of this stage, surface sinkholes enlarge by dissolution until some of them join, and the roof of the cave progressively collapses and is eroded away.

Vertical Shafts

LATE IN THE HISTORY of a limestone cave, after the water table has been lowered and air has largely replaced the ground water, some sculpturing may take place. It is during this period that a surface stream may enter the cave, carve notches on its walls, and leave a record of scallops and stream sediment. During this stage too, underground vertical shafts may be cut through the horizontal cave passages. Such vertical shafts are often called domepits because from below you look up at a dome, and from above you look down into a pit.

Vertical shafts are usually between 1 to 10 meters in diameter, with vertical extents of up to 50 meters. Their walls are characterized by vertical grooves, which contrast sharply with the smoother walls in most other parts of caves. Unlike most horizontal cave passages, which bear little relation to surface topography, vertical shafts generally lie under the heads of stream valleys, beneath the centers of sinkholes, or along a line where a layer of impermeable noncarbonate rock cover has been eroded off the limestone, for example, along a valley wall.

Cross section of Mammoth Cave, Kentucky. The oldest cave passages (at the top) and the youngest (at the bottom) formed at successively lower levels of the water table and Green River. Where the sandstone caprock is eroded off and surface water can enter the system, young vertical shafts cut through the horizontal cave passages. Moist passages under breached caprock contain calcite speleothems, whereas dry passages under intact caprock contain gypsum speleothems. The hundreds of kilometers of dry passages under the caprock are called the "Great Kentucky Desert." (*After* Thomas Poulson.)

Water often showers down inside a vertical shaft. Even casual observers get the impression that these features are younger than the main cave system, and that their formation often has no relation to the position of the older main cave passages. Commonly, only a narrow slot connects a vertical shaft with an older main passage. And vertical shafts often extend to the present water table although the main cave passages may be high above it.

E.R. Pohl has shown that vertical shafts form where cracks or joints have been enlarged through dissolution by water that moves rapidly from near the surface down toward the water table. The distinctive vertical grooves on the walls of the shafts are caused by a film of water flowing down the walls. The water retains a modest carbon dioxide content, usually acquired during lateral flow in noncarbonate material, and it has not been in contact with limestone long enough to exhaust its capacity to dissolve. Most vertical shafts are still being formed today by dissolution under aerated conditions, a process quite different from the underwater dissolution at or below the water table that forms the main passages of most caves.

Caves Formed by Hydrogen Sulfide Gas

A MOST SIGNIFICANT DEVELOPMENT in speleology is the discovery that some of the world's largest caves were formed by acidic solutions not from soil carbon dioxide, but from oil-field hydrogen sulfide. Hydrogen sulfide is familiar through its characteristic rotten-egg smell. This gas and its close relative, sulfuric acid, had long been known to play a role in some cave processes, but we now know that it was also responsible for the dissolutional activity that formed, among other caves, Carlsbad Cavern and Lechugilla Cave in Carlsbad Caverns National Park.

The enormous size and regularity of the rooms in Carlsbad Cavern had always been puzzling. Then trace amounts of native sulfur found there and in nearby Cottonwood Cave led Donald Davis to postulate that hydrogen sulfide had entered the caves and hence that sulfuric acid derived from it may have dissolved the limestone. Later, Carol Hill noted that the rare waxy clay mineral endellite, which William Davies and George Moore had discovered earlier in Carlsbad Cavern, requires a low pH, as would be provided by a strong acid such as sulfuric acid.

When organisms take energy from their food, electrons are a byproduct. Many organisms use oxygen to dispose of those excess electrons. In chemical terms, they *oxidize* the food material and *reduce* the oxygen.

But no free oxygen is available to oil-field bacteria. Therefore, to gain energy from "burning" petroleum, they must dispose of electrons elsewhere. The oil-field bacteria have developed the ability to add the excess electrons to sulfur in the mineral gypsum. They take energy from the petroleum for their own use, and in the process they also add energy from the petroleum to the sulfur in the gypsum. This reaction produces hydrogen sulfide gas.

One oil-field substance that the microorganisms use is the gas methane, which is the simplest hydrocarbon compound. They use it to convert gypsum to hydrogen sulfide by the following reaction:

$$\underset{\text{methane}}{CH_4} + \underset{\text{gypsum}}{CaSO_4 \cdot 2H_2O} \rightarrow \underset{\text{hydrogen sulfide}}{H_2S \uparrow} + \underset{\text{calcite}}{CaCO_3} + \underset{\text{water}}{3H_2O}$$

Hydrogen sulfide, being a gas, can migrate away from the oil field, where it then can play a role in cave origin. The gas moves upward into water-saturated limestone and finds its way up to the water table, where for the first time it meets with air—and with the oxygen contained in the air. The highly reactive hydrogen sulfide combines with the oxygen to produce sulfuric acid, which in turn dissolves the calcite of limestone to produce a cavity as follows:

$$\underset{\text{hydrogen sulfide}}{H_2S} + \underset{\text{oxygen}}{2O_2} \downarrow$$

$$\underset{\text{sulfuric acid}}{H_2SO_4} + \underset{\text{calcite}}{CaCO_3} \rightarrow \underset{\text{calcium sulfate solution}}{Ca^{2+} + SO_4^{2-}} + \underset{\text{carbon dioxide}}{CO_2} + \underset{\text{water}}{H_2O}$$

The carbon dioxide produced as a byproduct of this dissolution process can form carbonic acid that dissolves even more of the limestone. If the supply of oxygen is restricted, as it can be at some distance below the water table, part of the hydrogen sulfide is converted into yellow crystals of sulfur. Speleologists have now discovered tons of native sulfur in Lechuguilla Cave.

Hydrogen sulfide plays another important role in caves. In some caves it feeds microorganisms that serve as the beginning of a food chain that requires no input of energy from the surface. In a later chapter, we describe the best-studied example of this phenomenon, in Movile Cave, Romania.

Dating Caves and Cave Deposits

THE RADIOCARBON-DATING METHOD is one way to learn the age of stalagmites and related deposits that have grown during the past 40,000 years. Ordinary carbon in plants and animals contains a small percentage of carbon-14, a radioactive isotope generated by cosmic rays from nitrogen in the atmosphere. This radioactive carbon has a fixed concentration that is maintained in all living things, but when an organism dies, its radiocarbon decays back into nitrogen at a known rate.

Carbon dioxide produced by the decomposition of plant matter in the soil contains the same amount of radiocarbon as the plant matter does. Carbonic acid formed from this carbon dioxide dissolves the limestone. None of the carbon in the limestone is carbon-14, however, because the limestone is old enough for all the carbon-14 to have disappeared through radioactive decay. Therefore, the radiocarbon content of a saturated mixture of the limestone and carbonic acid from the soil would be 50 percent that of the original soil humus. But since the water is far from saturated with calcium carbonate by the time it reaches the cave, it contains an excess of soil carbonic acid, which is rich in radiocarbon. As a result, stalactites now growing have approximately 90 percent, rather than 50 percent, of the radiocarbon content of living plants. Knowledge of this fact permits us to date speleothems by the

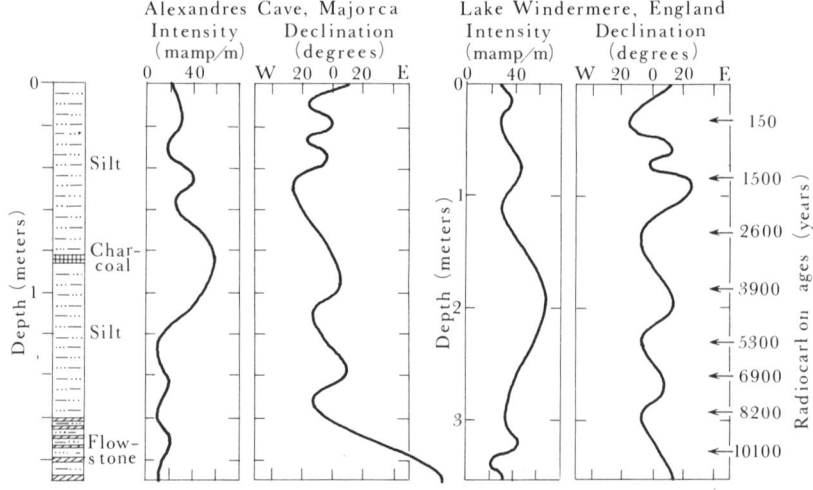

Magnetic minerals of cave silt align themselves with the Earth's magnetic field, which varies through time. Paleomagnetic curve matching with radiocarbon-dated lake sediment indicates the age of the cave silt. Flowstone near the base of this deposit in Alexandres Cave on the Mediterranean island of Majorca marks the end of the last glacial epoch, and charcoal near the middle records human intrusion into the cave. (*After* J.S. Kopper and Kenneth Creer.)

radiocarbon method, just as archeologists date the charcoal from ancient campfires.

By analyzing several samples from a cross section of a stalagmite from Moaning Cave, California, Wallace Broecker and his colleagues were able to date its rate of growth. They found that this particular stalagmite grew at a rate of 8.8 centimeters in 1,400 years—or about 0.06 millimeters a year.

Paleomagnetism of Cave Silt

SPELEOLOGISTS can use paleomagnetic measurements to date cave sediment in which carbon-bearing material suitable for radiocarbon analysis has been leached out or is too old. This method depends on the fact that the magnetic field of the Earth slowly fluctuates in direction and intensity. Sediment samples contain magnetic minerals that record the Earth's magnetism at the time the sediment was laid down. Magnetic measurements from European surface lakes of radiocarbon-dated samples in sediment cores show that during the past 10,000 years the position of the magnetic pole has migrated such that the direction to it has oscillated from one side to the other of 0° declination with a period of 2,700 years. At the same time that the position of the magnetic pole has fluctuated, the intensity of the field has also fluctuated.

Water-laid silt samples contain a remanent magnetization because magnetic mineral grains settle into alignment with the local magnetic field, or because new magnetic minerals grow with that alignment soon after deposition. The intensity, declination, and inclination of the fossil magnetic field are thus recorded down a vertical section through cave silt. Curves plotted for these magnetic properties in the cave are then matched with magnetic curves from a radiocarbon-calibrated section from a surface lake to interpret the age of the cave silt. Through a study of this kind, J.S. Kopper and K.M. Creer were able to date Magdalenian cave paintings in Tito Bustillo Cave, Spain, as being between 11,200 and 11,600 years old.

The measurements that track changes in compass direction toward the magnetic pole—the declination—can be followed with certainty for only a few 2,700-year orbits of the magnetic pole around the Earth's rotation pole. But another feature of the Earth's magnetic field that can be used for long-term measurement is the fact that it reverses completely approximately every 1 million years. During such a reversal, the north-seeking tip of a compass needle turns to point toward the south magnetic pole. This feature is particularly useful in providing a rough age for individual passages of large caves that typically are several million years old.

The last magnetic reversal was 780,000 years ago. Most large caves contain some passages older than that, so the line between the youngest passages with normal magnetization and any older passages with reversed magnetization is an important boundary for measuring the ages of various events in the cave.

The pattern of magnetic inclination at the surface of the Earth—horizontal at the Equator and vertical at the poles—shows that most of the magnetic field originates inside the Earth's iron core. Because iron has a lower melting temperature than the solid overlying rocks of the Earth's mantle, the outer part of the core is liquid, as is shown by earthquake waves passing through the core and received on the far side.

A natural dynamo in the liquid outer core causes the magnetic field. According to present theory, the magnetic field requires (1) the Earth's rotation, (2) the electrically conducting iron core, and (3) a slow overturn of the liquid outer core that moves electrons relative to the Earth's rotation. The movement of the liquid iron and its contained electrons starts a small magnetic field, either normal or reversed. This field, formed by the Earth's rotation, then generates an electric current, which in turn produces a stronger magnetic field, and on up to the full field.

If the average flow of the liquid iron and its contained electrons passes through an alignment parallel with the Earth's rotation axis, the magnetic field will collapse and then reestablish itself as a reversed field.

Cave samples for paleomagnetism are 2-cm cubes or cylinders of silt, carefully encased in plastic holders without disturbing their orientation, which is marked on the holders in the cave. In the laboratory, a sample goes into a room-temperature chamber in a magnetometer consisting of three mutually perpendicular superconducting coils in liquid nitrogen. The movement of the sample into the chamber generates currents in the coils. Once the specimen is in place, the currents flow continuously because of the superconductivity. The machine then reads out the properties of the magnetic field with respect to the specimen's orientation in the cave.

Victor Schmidt used the paleomagnetic method to show that the oldest passages in Mammoth Cave, Kentucky, are about 2 million years old. The four main levels of the cave range vertically through about 90 meters. We infer that the highest passages are the oldest, and that the lowest, at the level of the Green River, the youngest, because the river has cut downward while the various levels formed near the water table.

Paleomagnetism of Cave Silt

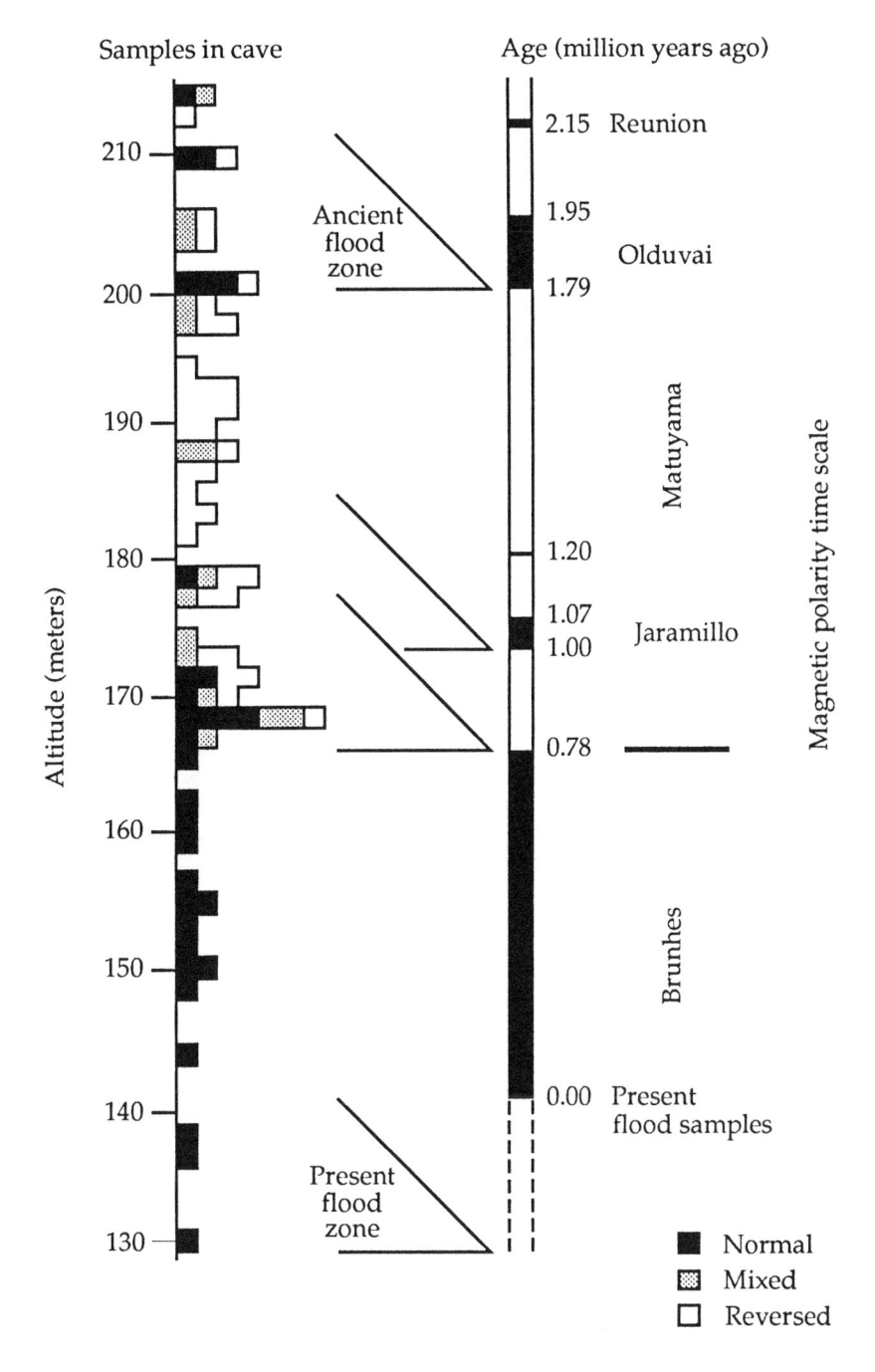

Silt samples at various levels in Mammoth Cave retain a fossil magnetization that is normal (black), mixed (gray), and reversed (white). The lower young parts of the cave contain silt magnetized only during the present normal period of magnetization (Brunhes Epoch). Upper parts of the cave contain mixed normal and reversed silt formed before 0.78 million years ago. (*After* Victor Schmidt.)

The silt up to somewhat less than half the full vertical range of the cave is normally magnetized, whereas the upper passages are mixed normal and reversed. Inasmuch as the last reversal was 0.78 million years ago, if the river downcutting was approximately constant in long-term average, then the data imply that the highest levels of the cave are more than twice as old as the last reversal.

F.-D. Miotke and A.N. Palmer argue that stream terraces along the Green River show that the four levels in Mammoth Cave were formed successively during four Pleistocene glacial stages.

Patches of cave silt in most caves are discontinuous, because both flood deposition and cave-stream erosion are irregular. Hence identifying which normal or reversed magnetic interval a particular patch of cave silt represents can pose a problem. Charles Repenning has used the fossil teeth of meadow mice to help resolve this problem.

Meadow mice—or voles—are small mammals that live in temperate regions of the Northern Hemisphere. Their preserved remains, especially their tiny teeth, are often found throughout cave systems, although they themselves are not cave animals. Evidence suggests that owls bring their remains to cave entrances and deposit them there in owl pellets. Cave rats then pick up the pellets and bones and carry them throughout the cave.

Meadow mice evolved rapidly during the past 7 million years, the last 2 of which coincided both with the waxing and waning of the ice-age glaciers and with the development of most limestone caves. The mice reach maturity in only 20 days, and they breed year-round. One mouse can produce millions of offspring in just a few years.

Their range extends to the Arctic tundra, and during low stands of sea level they moved freely between North America and Asia and Europe. Their convoluted narrow and tall teeth, which are a little larger than very course sand grains (1 x 3 x 3 mm), evolved continuously. Today the teeth serve as markers both for the evolution of the animals and for the age of the caves.

During periods of high sea level, when the ice-age glaciers melted and their water returned to the oceans, meadow mice could no longer migrate across Bering Strait. This prevented their intercontinental breeding, and the evolution of species ran separately in the two hemispheres. During later periods of low sea level, while glacial ice returned to the continents, the new and distinctive species crossed the exposed continental shelves, and their teeth appeared abruptly as markers in cave silt.

Now we need merely to identify the assemblage of meadow-mice species in a sample of cave silt, combine it with the magnetic polarity

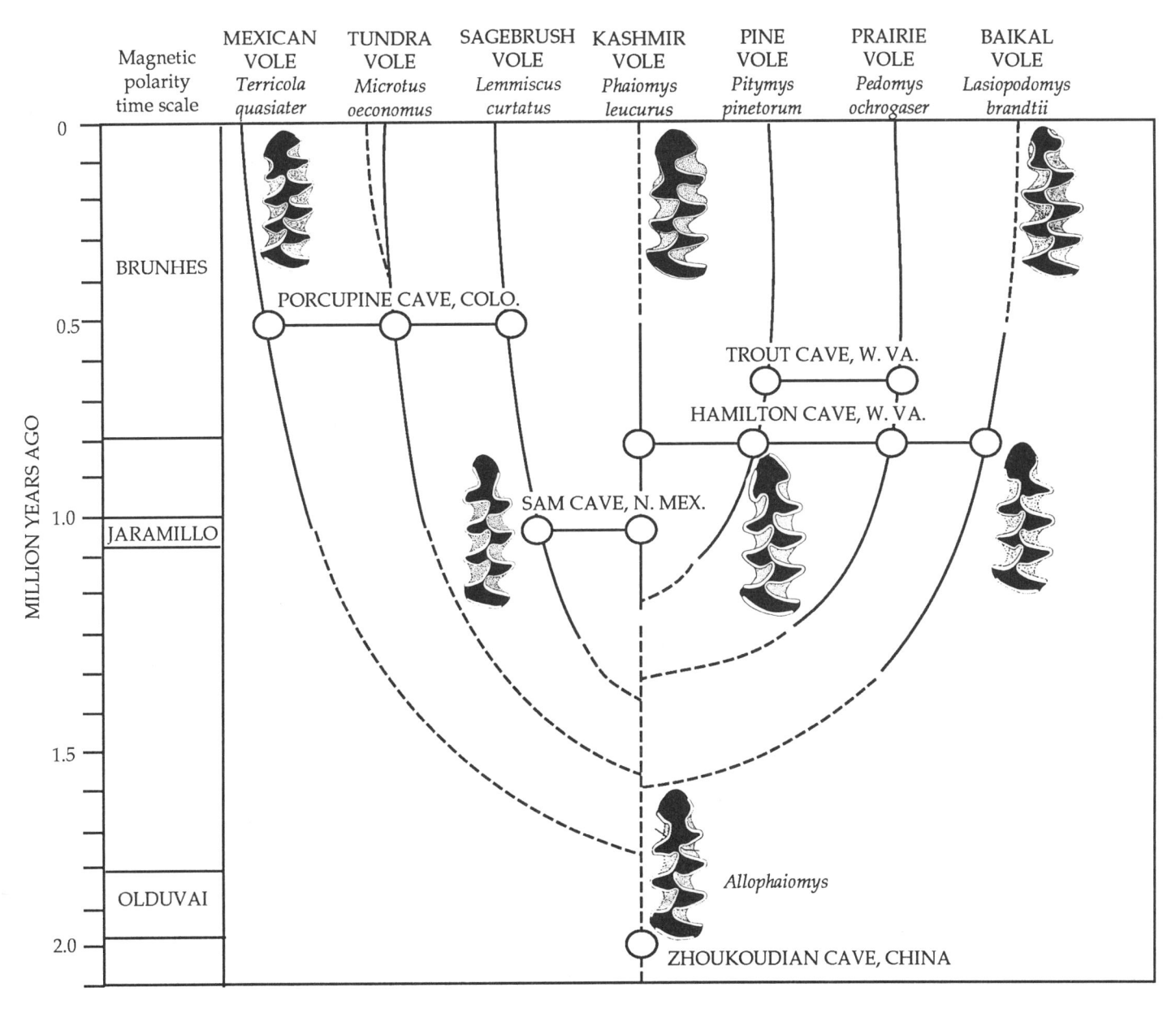

Meadow mice (voles) evolved rapidly during the past 2 million years, and their distinctively marked teeth are preserved in many Old World (dashed lines) and New World caves. They moved freely across Bering Strait during ice-age glaciations, when sea level was low, but the high sea level of interglaciations prevented interbreeding. Their fossil teeth calibrate times when the polarity of the Earth's magnetic field as recorded in cave silt was normal or reversed. (*After* Charles Repenning.)

of the silt, and we obtain a good estimate of the time when that cave passage was open.

DURING 1995, Josep Parés and Alfredo Pérez-Gonsález used paleomagnetism to date the oldest human fossils in Europe. A railroad cut at Gran Dolina in northern Spain intersected a sediment-filled pit cave that had trapped human remains and other material. About 18 meters of strata consisting of silt, limestone fragments, and thin stalagmitic and bat-guano layers fill the cave. Artifacts and bones that belong to ancestors of the Neanderthals lie at about the middle of the sediment fill. Paleomagnetic measurements show that the last magnetic reversal—the boundary between the present normal magnetic period (the Bhrunes) and the preceding reversed period (the Matuyama)—occurs near the human fossils. Hence these remains are about 780,000 years old, an age that is about a quarter of a million years older than was previously thought for the migration of the earliest humans into Europe.

Karst

ROCK DISSOLUTION, which is so important to the origin of limestone caves, also affects the land surface in cave areas, and this leads to a distinctive topography known as karst. The word *karst* is derived from Slovenian, either from *kar* (rocky) or *hrast* (oak), and was first used by Austrian map makers in 1744 as a name for the rocky, oak-forested karst and cave region that lies in Slovenia and northeastern Italy. The word has since been generalized to mean any terrain where the topography has been formed chiefly by the dissolving of rock.

Karst areas commonly have inward-sloping surface depressions, and the drainage is subterranean through caves. Bare rock in karst regions is usually covered by U-shaped dissolution grooves or channels, from about 1 millimeter to 1 meter across, separated by sharp ridges. Where the rock is covered by soil, or where the soil has recently been removed, there are often rounded fissures along joints.

Karst is characterized by several different types of large-scale topographic features that are gradational with one another. In cave areas of the humid-temperate parts of the United States and Europe, the usual type is *sinkhole karst*, characterized by funnel-shaped depressions. These sinkholes have dimensions most commonly measured in meters or tens of meters. The upper level of sinkhole karst forms an otherwise con-

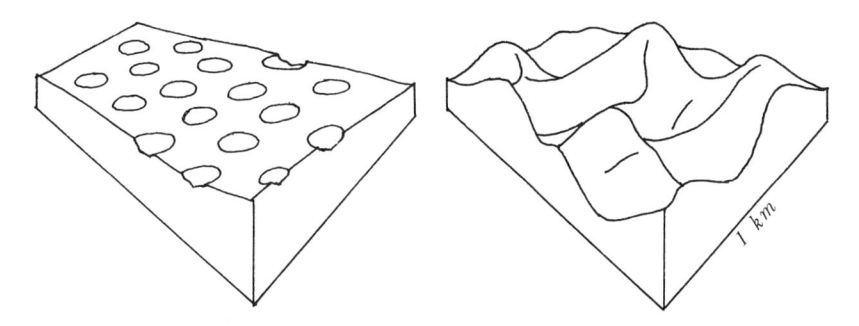

Temperate-karst sinkholes (left) average about one-tenth the diameter of tropical-karst cockpits (right). The cockpit karst requires an average temperature of at least 18°C, a rainfall of 1200 millimeters, and rapid mountain uplift.

tinuous surface, interrupted by the sinkholes, which in many places impinge against one another.

Sinkholes are a costly natural hazard in limestone areas. W.E. Sinclair has shown that when pumps lower the water table beneath sinkholes, fast-moving streams form underground. These streams remove unconsolidated sediment that previously supported the surface, and the sinkholes may suddenly collapse and swallow up buildings.

Humid-tropical cave areas, such as Puerto Rico and southeast Asia, are distinguished by two related types of karst. The first, *cockpit karst*, consists of a terrain made up of conical hills alternating with polygonal or star-shaped depressions, the whole resembling a molded egg box. The second, *tower karst*, consists of more widely separated steep-walled hills rising from flat valleys or plains. Such tower-like hills with rocky overhanging cliffs usually have caves at their bases, and have often been depicted in the art of southern China.

The lowland flats in the tower karst of tropical regions seem to be related to large basins in temperate-karst regions known as *interior valleys*. An interior valley is an enclosed depression several kilometers in diameter that has a flat floor and steep walls. During the rainy season, springs feed streams that cross the fertile floor of the valley and sink into caves on the lowest side. During especially heavy rains the caves may not be able to handle the runoff, and within a few hours the interior valley becomes a lake. Later, at the end of the rains, the lake usually disappears as quickly as it formed.

The presence of soil, and its nature and distribution in different climatic zones, is believed to be an important factor in the development of the various types of karst. The cliffed walls of karst towers and of interior valleys are clearly related to rapid dissolution of limestone under a soil blanket in the valley and at the foot of the cliffs. This soil blanket forms a cover that holds in carbon dioxide that combines with surface water to form carbonic acid that speeds the process of dissolution.

The development of karst features requires that the process of dissolution be faster than other types of erosion that affect all rocks. Where moisture is sparse or temperatures usually low, the distinctive karst landforms develop only poorly. Hence, in cold and arid lands, where dissolution is slow, limestone underlies mountains, whereas in warm and humid areas, where dissolution is rapid, limestone underlies lower topography than other types of rock nearby.

The speed of erosion in typical karst areas is such that karst tends to be ephemeral, because limestone in humid climates tends to erode rapidly to base level. As a consequence, the great karst and cave areas of the world are in two settings: regions of rapid mountain uplift, and regions where limestone beds are being exposed by the erosional retreat of a nonsoluble rock cover.

Caves not in Limestone

THREE TYPES of caves besides dissolution caves are important to speleology—sandstone caves, sea caves, and lava tubes.

Sandstone caves were widely used for shelter by early peoples, who liked them all the better for their being shallow. Deep limestone caves were too wet, cold, and dark for comfortable habitation. Sandstone caves are formed at the base of cliffs, where certain parts of the rock are less well cemented than other parts. Surface water moving down the cliff dissolves away the cement in those areas, so the sand grains fall to the ground and are washed away. At the same time, upper sandstone surfaces get harder because additional cement is deposited there from water drawn to the surface by capillarity. Good examples of sandstone shelter caves are those that contain the famous cliff dwellings at Mesa Verde National Park in Colorado.

Sea caves are somewhat similar to sandstone shelter caves, except that they occur along the shore. They are formed at the base of sea cliffs at places where one part of the rock is more easily eroded than the adjacent rock, allowing wave attack to remove it more easily. Sea caves are usually formed along vertical zones of weakness, such as faults or steeply dipping beds of softer rock. Because the waves can attack only the base of the cliff, the weak zone retreats faster there than it does higher up; the higher part of the cliff therefore remains as a roof for the cave. The Pacific coast of the United States has an especially large number of sea caves, some of which, like Sea Lion Cave in Oregon, are tourist attractions.

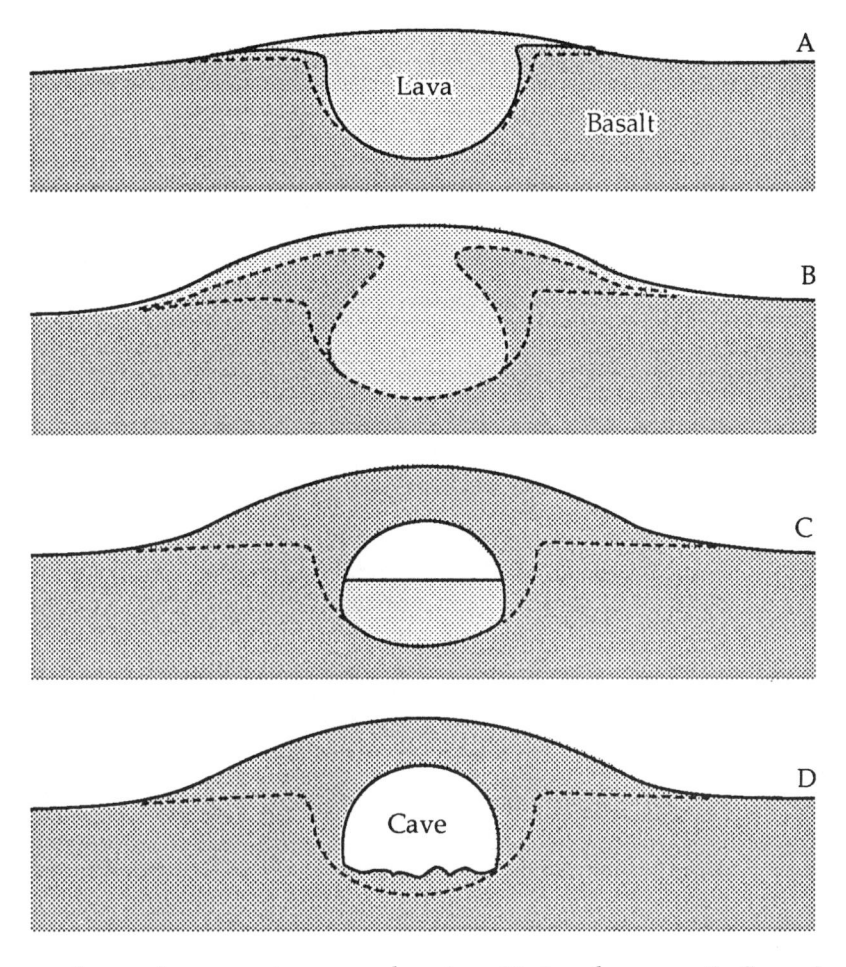

Successive stages in roof development of a lava cave. (A) Repeated overflow and subsidence of the lava stream build a levee and start inward growth of the top of the channel; (B) fluctuation of the level of active lava builds the banks upward and inward and causes the lip to thicken; (C) the roof closes over, and continuing variation in flow causes the ceiling to thicken; (D) the finished cave has an irregular floor from the last stages of lava movement. (*After* Donald Peterson and others.)

Lava tubes occur in areas where basaltic lava has recently flowed from volcanoes. A tube forms when a tongue of lava, flowing down a marked slope, solidifies on its outer surface, while the interior remains molten and continues to flow. When the liquid lava has drained out of the interior of the tongue, a tubular cavity remains. A newly formed lava cave does not have an entrance at the level of the floor, but a thin part of the ceiling commonly collapses after the lava has cooled, and thus forms an opening through which the cave can be entered. Lava caves can be many kilometers long and can branch in both the upstream and downstream directions of lava flow. Almost all areas of young basaltic rocks contain some lava caves, and many of those in the western United States and Hawaii are visited by tourists.

When many measurements of air currents must be taken simultaneously in different parts of a cave, sheets of aluminum foil serve as fairly precise anemometers. The deflection of the bottom of the foil being used here in Breathing Cave, Virginia, is measured with a meter stick.

3: Characteristics of the Underground Atmosphere

MANY of the familiar roadside signs advertising caves open to the public carry the phrase, "Come underground and cool off." This is reasonable advice, for during the heat of summer the temperature of these caves is indeed pleasantly cool. If the principal tourist season were in the winter instead of the summer, however, the signs would probably read, "Come underground and warm up," for in winter these same caves are far warmer than the surface. Cave temperatures are nearly constant throughout the year.

Controls of Cave Temperature

THE AIR MOVEMENT in and out of a cave is so slow that within a few hundred meters from the entrance, the cave air ordinarily takes on the same temperature as the wall rock. The temperature of the deep parts of limestone caves is therefore controlled by the temperature of the limestone, which is, in turn, approximately equal to the average annual temperature of the surface.

The daily and seasonal temperature fluctuations of the surface tend to diminish as the heat moves down through the rock into the cave. A 30°C temperature fluctuation between day and night is reduced to a fluctuation of less than 1°C at a depth of 57 centimeters below the surface. Likewise, a fluctuation throughout the year of 30°C is reduced to 1°C at a depth of 11 meters. Because caves commonly lie deeper than

11 meters, they ordinarily have annual temperature variations of less than 1°C.

The surface temperature on which the cave temperature depends is chiefly determined by the latitude of the cave and its altitude above sea level. The average temperature of caves near the southern border of the United States is about 20°C, and that near the northern border is about 5°C. The effect of altitude is illustrated by two Colorado caves near one another and at approximately the same latitude. Fly Cave, at an altitude of 1920 meters, has a temperature of 13°C, whereas Spanish Cave, at an altitude of 3630 meters, has a temperature of only 2°C. This indicates a gradient of about 6°C per thousand meters. The temperature of middle-latitude limestone caves for which the location and altitude are known can be approximated by the following equation:

$$°C = 38 - 0.6L - 0.002h$$

where L is the latitude in degrees and h is the altitude in meters.

Other factors besides the latitude and the altitude affect the temperature of many caves. Because water is more efficient than air in transporting heat, the temperature effect of a stream extends much deeper into a cave than that of air currents. Caves that receive large quantities of snowmelt water in the spring have much lower temperatures than normal for caves of that particular altitude and latitude.

Cave-temperature map of the United States in degrees Celsius. The temperature of a cave is approximately equal to the average annual temperature of the surface.

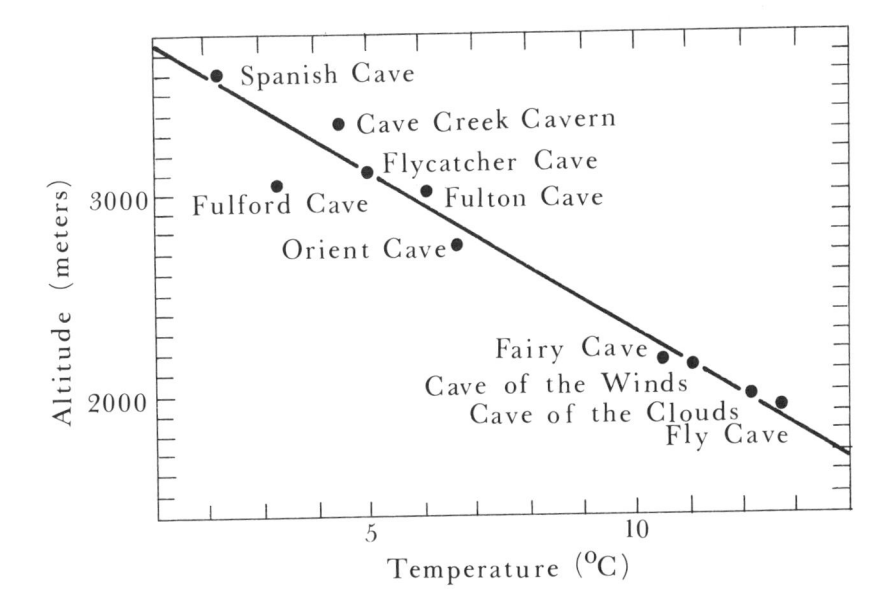

Effect of alitutde on the temperature of ten caves at about the same latitude in Colorado (*Unpublished data supplied by Donald Davis.*)

Caves under north-facing slopes would presumably be a little cooler than caves under sunny, south-facing slopes, but this temperature difference is small and has not yet been studied to any extent.

The rate at which an upper-level temperature change moves downward in limestone from one depth to another is proportional to the temperature difference between the successive depths. As this gradient becomes smaller and smaller with increasing depth, the time required for a change in surface temperature to affect the air in a cave greatly increases. At a depth of 11 meters, the effect of the winter's cold is felt in the following summer, but at greater depths the effect of a marked rise or fall in average surface temperature may not become apparent for hundreds or even thousands of years. The temperatures at these lower depths can thus be regarded in part as "fossil" temperatures.

A formula for approximating the depth in meters, x, at which a temperature change of $1°$ results in limestone from a cyclic surface temperature change lasting T years is

$$x = 3.18 \sqrt{T} \ln N$$

where $\ln N$ is the natural logarithm of the surface temperature fluctuation N in $°C$.

Temperature of Model Cave, Nevada, on August 27, 1952, plotted against depth. The statements regarding the meaning of parts of the curve are based on the assumption that all heat transfer occurred by conduction through the limestone. (*After* Arthur Lange.)

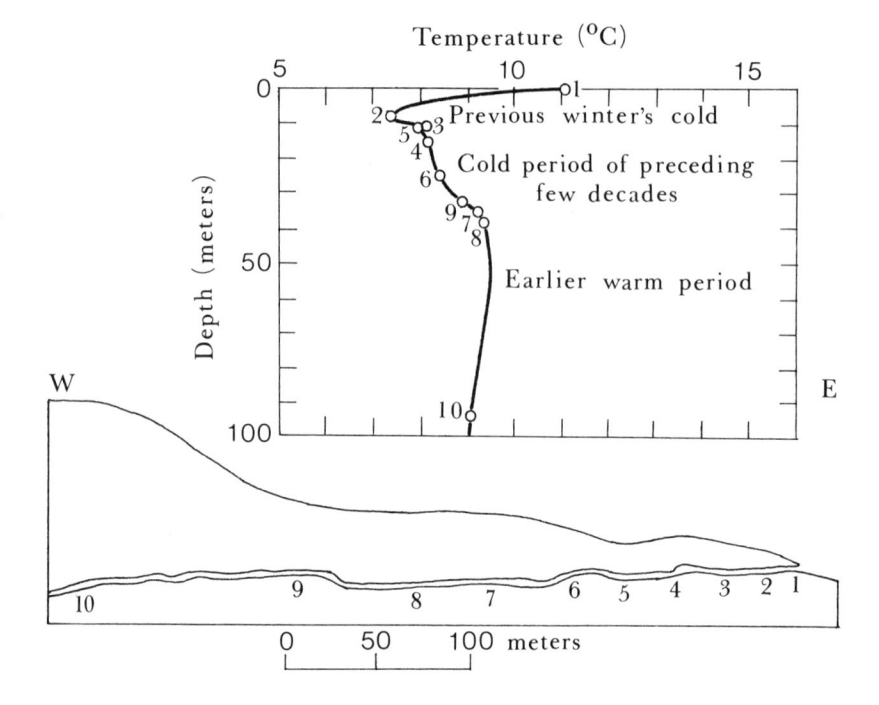

Probably the oldest climatic event that might conceivably be recorded in a cave is the last cold stage of the Pleistocene ice age. Assuming that the cycle of which it was a part began 40,000 years ago and that the average low temperature reached was 10°C lower than at present, the formula indicates that the temperature effect of this last glaciation would now equal 1°C at a depth of 1464 meters. The deepest cave so far discovered in the world, the Jean Bernard System, France, is 1602 meters deep. Thus, the temperature near the bottom of the Jean Bernard System presumably reflects the average annual temperature of the region as it was as long as 40,000 years ago.

Caves Containing Perpetual Ice

A NORTHERN CAVE that lies at high altitude will have a below-freezing temperature if the average annual temperature of the surface is less than 0°C. Such caves, because they commonly contain ice all year round, are called *ice caves*.

Very large crystals of ice can grow in ice caves. The floors of such caves are usually coated with transparent sheets of ice made up of hexagonal prisms 2 to 5 centimeters in diameter, standing vertically. The honeycomb-shaped boundaries between the crystals are clearly visible on the surface of the ice. Large platelike ice crystals with well-developed crystal faces may also grow on cave walls and ceilings. These crystals are usually formed by deposition directly from the atmosphere of the cave.

Some caves function as cold traps and may become ice caves even though the average surface temperature in their vicinity is above freezing. Cold-trap caves are generally bottle-shaped, that is, they open up downward from narrow entrances. In the winter, dense cold air flows into the cave and fills it from the bottom; in the summer, circulation ceases, and the cold air is trapped in the cave. Such caves may have temperatures nearly $10°C$ below what would be expected from their latitude and altitude.

Cold-trap caves in limestone are relatively rare, because limestone caves seldom have the required relationship between passages and entrances. Lava tubes, on the other hand, almost always have the necessary shapes, because their entrances are holes formed by the collapse of part of the ceiling. One can visit interesting groups of lava-tube ice caves of this type in Lava Beds National Monument, California, and Craters of the Moon National Monument, Idaho.

Relative Humidity

THE AIR OF MOST CAVES is nearly saturated with water vapor—in other words, the relative humidity is close to 100 percent. This is so because seeping water moistens the ceilings, walls, and floor that the air must pass by as it moves slowly through the cave. The constant temperature of the inner part of the cave permits this high humidity to be maintained indefinitely.

Near the entrances to caves, however, the humidity may be lower, partly because the outside humidity is usually lower, and partly because the cave temperature differs from the outside temperature. A fall in the temperature increases the relative humidity, and a rise in temperature decreases it. In the summer, warm air entering a cool cave soon becomes saturated without absorbing water from the cave walls. In the winter the air becomes warmer as it enters the cave, and for a short distance its relative humidity falls. Deeper in the cave, however, the humidity of this warmed air slowly rises to the saturation level.

The opposite effect occurs in caves from which air currents flow outward. In the winter, the saturated air extends all the way to the entrance. In the summer, the humidity may be low just inside the entrance where the outward-moving saturated air first becomes warmed.

Air Currents Caused by Barometric Change Near Entrances

THE PARTS OF CAVES near entrances are ventilated by an exchange of air with the outside. This exchange varies as a function of the constantly changing pressure of the outside atmosphere. Surface barometric pressure changes are of two types, periodic and nonperiodic. The most important periodic change is a 24-hour fluctuation resulting from the difference in the temperature of the air between day and night. During the day the air is warmed and becomes less dense, so the pressure falls. At night the air is cooled, and the pressure rises. Normally, then, the air will begin to flow into a cave at sunset and will begin to flow out at sunrise. Nonperiodic changes in barometric pressure are related to the weather, such as, for example, the pressure changes that accompany the passage of a storm front. These are superimposed upon the daily fluctuation, and the cave pressure adjusts itself to conform to the resultant effect of both.

Because a cave usually consists of a series of small, tortuous, interconnected passages with relatively small entrances, a considerable time may be required for the air moving through the cave to equalize cave air pressure with outside air pressure. The air in most caves is therefore almost constantly in motion, adjusting itself to the surface changes. Usually the cave air currents are so slow that they can be detected only with instruments. In constricted passages, or at small entrances to large caves, air currents caused by surface barometric changes are sometimes detectable as light breezes on one's face. Stronger cave winds, some of which are veritable gales, are generally caused not by barometric change but by a chimney effect.

Chimney and Reverse-Chimney Effects

IN SOME CAVES that have two entrances, one higher than the other, a fairly strong air current issues persistently from one entrance or the other. These caves are called *blowing caves*. They normally have an annual cycle, in which air blows out of the lower entrance all summer and out of the upper entrance all winter.

In the winter, a blowing cave functions in the same way as a chimney. Cold air enters the lower entrance, is warmed inside the cave because there the temperature equals the average annual temperature of the surface, and rises to emerge from the upper entrance. In the summer the movement is reversed. Because the air inside the cave is colder than the outside air, it flows out through the lower entrance, while air flows in at the upper entrance. The steady current of humid air issuing from a blowing cave often causes a lush growth of moss and ferns to form around the entrance. The humid air may also condense in the winter, forming a vapor column visible from a great distance.

Breathing Caves

IN A FEW CAVES, the air moves inward for a few minutes and then outward for a few minutes, as if the cave were breathing. Burton Faust studied this effect at a cave in Virginia now known as Breathing Cave. While waiting in a small passage near the entrance for the remainder of his party to come out, Faust noticed an unusual reversal of the air current in the passage and lit a candle to study the effect further. The

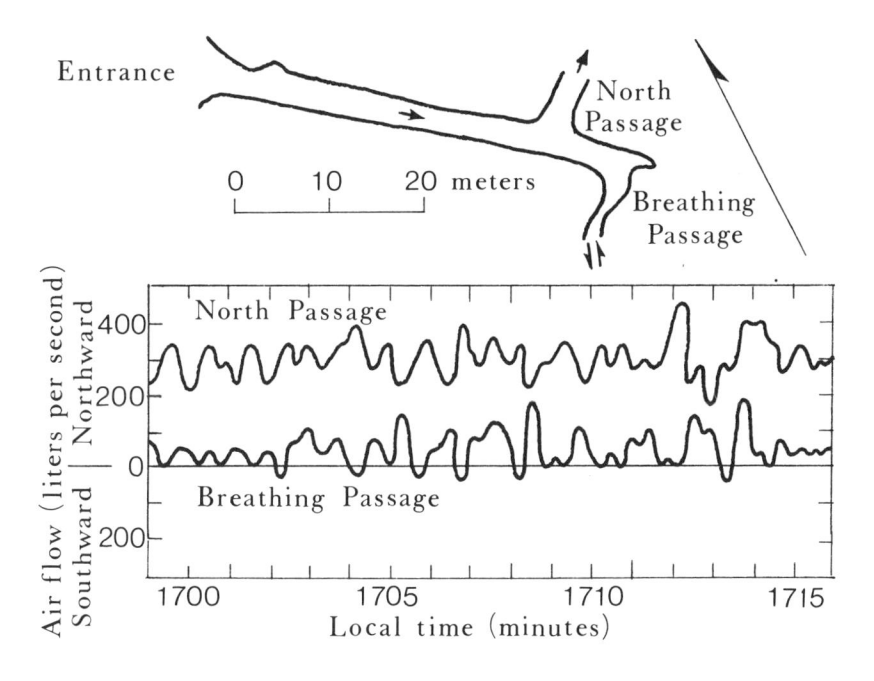

The breathing phenomenon in Breathing Cave, Virginia, on May 7, 1955. A flow of air, averaging 300 liters per second, moves from the entrance into North Passage, probably in response to a chimney effect. This continuous flow past Breathing Passage causes it to breathe by making the passage operate as the neck of a resonating chamber.

candle flame was first deflected by the moving air toward the interior of the cave and then, after standing upright for a moment, was deflected toward the entrance. This cycle subsequently repeated itself again and again.

The breathing is at the inner end of a straight passage about 45 meters long, known as the Entrance Passage. Inward from this point the cave continues as an extensive series of rooms. About 3 meters nearer the entrance, a lateral opening to the north leads to another series of rooms. In an effort to learn more about the phenomenon, we made a series of simultaneous observations of the air velocity in May 1955 in Entrance Passage, Breathing Passage, and North Passage. A somewhat irregular air current blew continuously from the entrance into North Passage, past the mouth of Breathing Passage. At the same time the air in Breathing Passage oscillated with a cycle about a minute long.

The breathing-cave phenomenon may be compared with that of a compound Helmholtz resonator. A familiar example of a simple Helmholtz resonator is a cider jug. When one blows across the mouth of the jug, one sets up vibrations that can be heard as a low musical note. The air in the jug acts as a spring, compressing and expanding with a certain resonant frequency. A breathing cave is rather like a gigantic, very irregular jug, but its great size makes the oscillation of the air mass inside the cave much too slow to produce an audible note. A formula for approximating the breathing cycle of a jug-shaped cave operating as a simple Helmholtz resonator is

$$T = 0.019 \sqrt{lV/S}$$

where T is the period of the breathing cycle in seconds, S is the cross-sectional area of the necklike breathing passage in square meters, l is the length of the passage in meters, and V is the volume of the cave beyond in cubic meters.

The complicated passages of Breathing Cave, Virginia, make it difficult to insert their dimensions into the formula for a simple Helmholtz resonator. Moreover, when we attempt to insert them, we find the breathing period predicted by the equation to be shorter than the actual breathing period. This suggests that the simple jug model does not apply to this cave, and that we must calculate the period of a compound Helmholtz resonator rather than that of a simple one. The period of resonance in a compound resonator, in which several large air chambers are connected by small tubes, is lower than that of a simple resonator of the same volume.

The triggering agent for some breathing caves is probably the turbulence in the wind blowing past the entrance. The velocity is less important, for the resonator responds to minor disturbances in the airstream that are in phase with the resonant frequency.

Breathing Cave seems to be much more favorably arranged for this phenomenon than a cave that depends on random wind movement across its entrance. The irregular but constantly inward movement of the air into North Passage in May suggests that this passage serves as the beginning of a chimney in which cool air moves into the north part of the cave, is warmed, and goes out through a higher exit. Usually, therefore, the air moves between the entrance and North Passage, past the mouth of Breathing Passage, either inward or outward depending on the season.

The arrangement of the junction is like that of the mouth of a whistle or an organ pipe. The steady breath of air in an organ pipe provides vibratory energy when the lip of the pipe mouth deflects it first one way, then the other. The air current provides the energy, but the resonance of the pipe determines the frequency. Precisely the same thing happens in Breathing Cave, except that the dimensions are so large that the oscillation, instead of being measured in cycles per second, must be measured in cycles per hour.

Ebb and Flow Springs

THE WATER FLOW from some large springs in limestone regions exhibits a curious pulsating action. Every few hours a surge of water about 10 times the base flow erupts from the spring. The process that triggers this ebb and flow is not yet definitely known, but enough has been learned about the phenomenon to ascribe it tentatively to an intermittent siphon.

In most cases, study of ebb and flow systems necessarily has been restricted to measurements of changes in the flow of outside springs, but in research begun under the direction of Stanley Ulfeldt at Big Spring, California, effects in Lilburn Cave, 700 meters away from Big Spring and about 10 meters higher, have also been observed and compared with those taking place simultaneously outside.

In early phases of the study at Lilburn Cave, speleologists inside the cave clocked a slow rise of ponded water succeeded by a sudden flushing, while others at Big Spring noted a steady increase in flow followed by an abrupt flood.

3 / Characteristics of the Underground Atmosphere

Fluctuation of water in Lilburn Cave, California, and at Big Spring, 700 meters away. During these ebb and flow cycles, which are believed to be caused by an intermittent siphon, the delay time between a cave flush and its corresponding flow at the spring is 20 seconds. (*Unpublished data supplied by* Stanley Ulfeldt.)

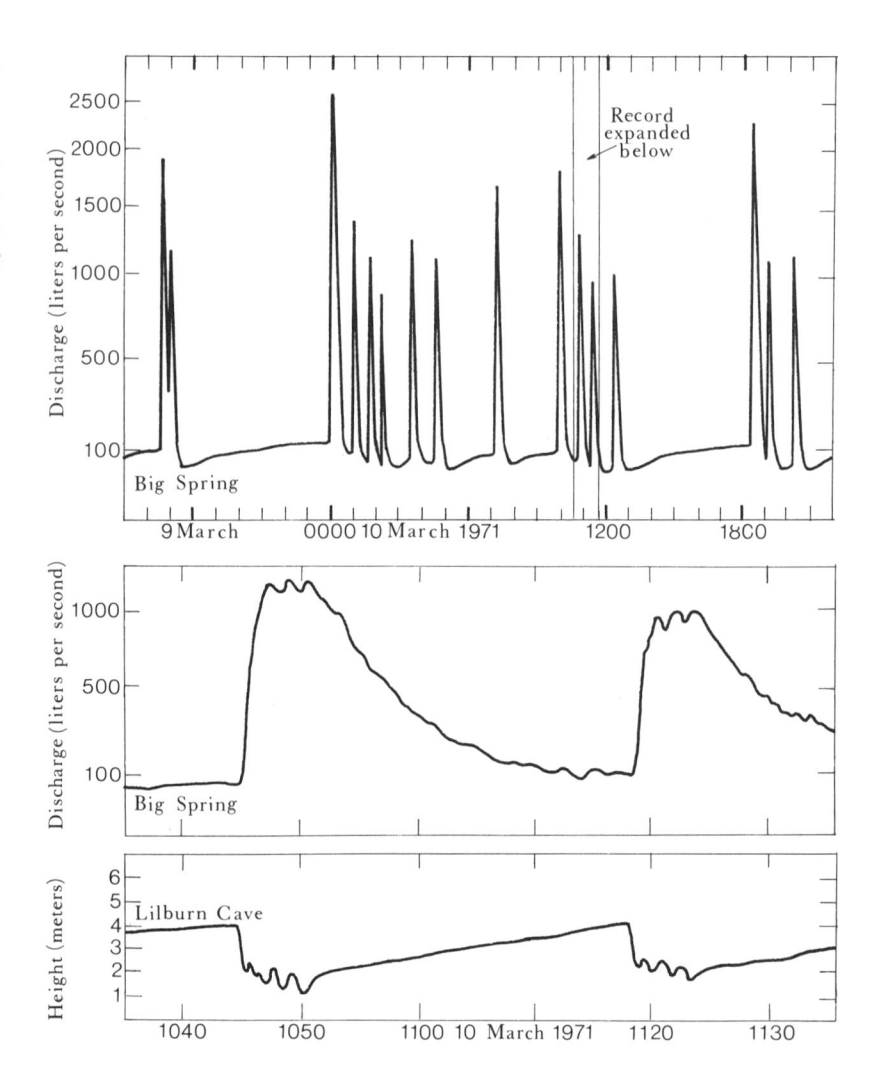

To extend the period of observation and to increase the precision of the time correlation between events at Big Spring and at Lilburn Cave, Ulfeldt set up a permanent water-level gauge at the spring, and for special observation periods installed pressure transducers in the cave pool and at the spring, both recording electronically on a single chart.

The recordings show that the ebb and flow action is confined to periods of moderately high runoff from snowpack and from rains. The sudden discharges commonly occur in groups in which each successive flood has a lesser flow than the previous one. Both the number of floods in a group and the quantity of water released in the first flood are clearly dependent on the length of time elapsed since the end of the last group of floods. An especially long period without floods is followed by an abnormally large surge and also by a greater than average number of subsequent surges.

In addition to the major flushing action, the detailed recordings, both in the cave and at the spring, show oscillations that have a period of about 100 seconds. These oscillations are almost exactly out of phase with each other at the two sites, suggesting that the cave and spring are part of a u-tube oscillating system, in which the water moves from end to end with a characteristic frequency. The identical period of the oscillations indicates that during the flushing, a water connection between the cave and spring is continuous. The oscillations end abruptly in the cave after the flushing, when the water resumes its slow rise.

Flushing may occur in the system when water enters the cave too fast to be handled by spillover at the apex of a siphon passage. The water then rises in the cave until it impinges against the ceiling at the apex, a siphon action is established, and the water in the cave reservoir is rapidly drawn down to a level where air again enters the system from the air-filled part of the cave and breaks the siphon.

The intervals between the surges might be related to variations in the height of the u-tube oscillations. The maximum amplitude of these is about 25 percent of the total water-level fluctuation in the cave. When the oscillations are high, the water may impinge against the ceiling early and establish a siphon more quickly, thereby reducing the time interval between surges.

This hypothesis provides an explanation of the occurrence of surges in groups. After a long steady period of water rise in the cave, presumably with little oscillation, a large surge finally occurs and sets in motion a train of strong oscillations. At the end of this first surge, the siphon breaks, but this happens before all the water in innumerable pas-

Cross section of an ebb and flow spring system. (A) The stream from the cave to the spring is at base flow; (B) the flow exceeds the capacity of the system, the water rises in the cave, and air at the top of a siphon passage is entrained by turbulent water and swept out at the spring; (C) the increased head at the cave initiates a siphon action that draws down the water in the cave, causes a surge at the spring, and then admits air to break the siphon; (D) base flow is reestablished.

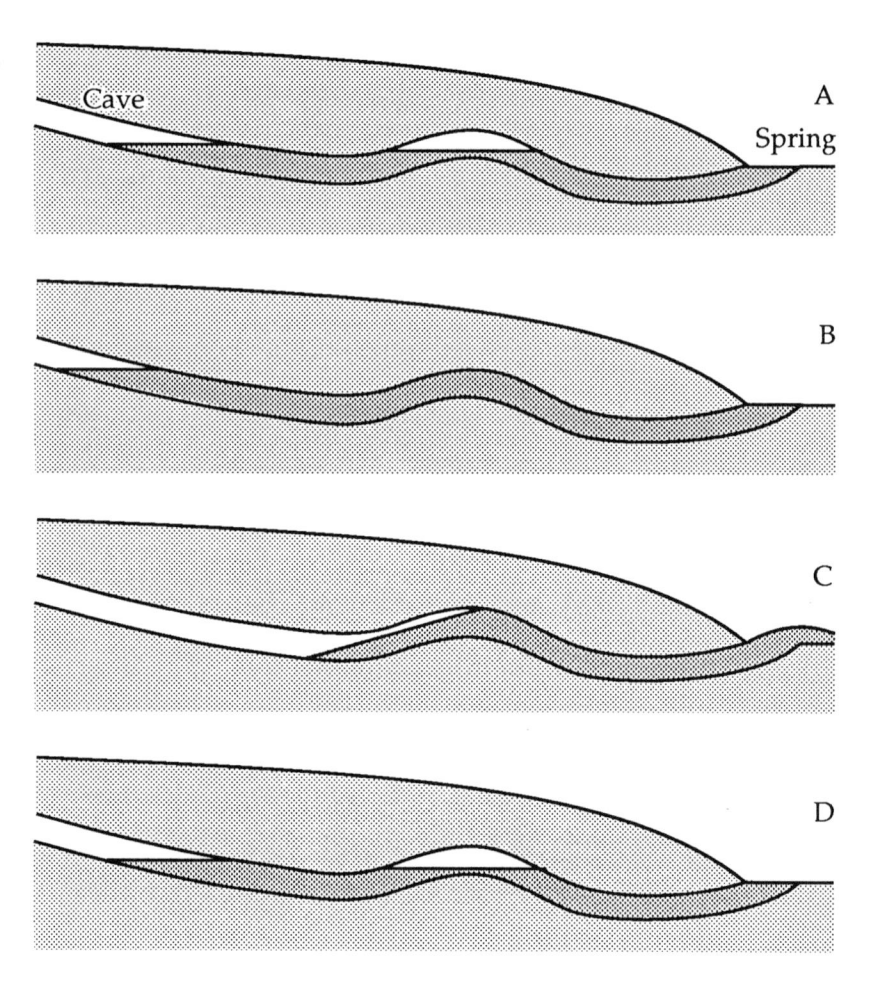

sages that constitute the upper reservoir can drain out. Hence, a potential for subsequent surges remains in the still amply full upper reservoir, and these are triggered by the persisting oscillations in the u-tube between the siphon and the spring. Eventually, the water in the upper reservoir is depleted, the oscillations die out, and a period of recharge passes before the next group of surges takes place.

This type of ebb and flow action is not unique to the system at Lilburn Cave, California. At Ebb and Flow Spring, Missouri, an almost identical relation is observed between the length of time preceding a group of surges and both the surge height and the number of surges in the group.

A column in Whipple Cave, Nevada, formed when a stalactite growing downward from the ceiling met a stalagmite growing upward from the floor.

4: Growth of Stalactites and Other Speleothems

EACH YEAR thousands of visitors descend from the surface of the Earth into the mysterious dark world of caves. Much of the appeal of caves arises from a feeling of unreality imparted by the darkness and the silence. But what attracts people most strongly to this netherworld is the strange underground landscape, forested with stalactites and stalagmites, and decorated also in places with hanging curtains of stone, massive richly sculptured columns, and delicate, flowerlike structures. Most people who see these things wonder how they were formed. Here we answer a few of the questions asked about them.

Features formed in caves by the deposition of secondary minerals—such things as stalactites and stalagmites, for example—are commonly referred to as "formations." But because geologists use the word *formation* in a very different sense, it is better to call these things by a general term that applies only to them, and such a term that has come into general use is *speleothem*. This word is derived from the Greek words *spelaion* (cave) and *thema* (deposit). The term is applied only to deposits formed from a chemical solution or by the solidification of a fluid after the formation of a cave. Stalactites are speleothems, as are ice crystals formed directly from water vapor in a cave; but beds of silt, and calcite veins etched into relief when the walls of the cave were dissolved, are not.

The history of the study of speleothems is long, but relatively few investigators have employed the modern tools now available to mineralogists. A scanning electron microscope displays the microstructure of speleothems, a polarizing microscope reveals their crystal structure, chemical analyses give their compositions, and x-ray diffraction discloses the presence of minerals that may not be identifiable under the microscope.

Importance of Soil Carbon Dioxide

SPELEOTHEMS consist mainly of calcite, the same mineral that makes up the wall rock of limestone caves. The calcite is dissolved from limestone above the cave by slightly acidic water percolating downward from the soil zone. When it reaches the interior of the cave, this water deposits the calcite it holds in solution, thus forming a stalactite or other speleothem.

Everyone knows how to make salt crystallize by evaporating water from a brine. But since the air in a cave is saturated with water vapor, none of the water that enters can evaporate. What makes the dissolved calcite come out of solution is a totally different process—the loss of carbon dioxide gas from the dripping water.

In the soil zone, where abundant plant remains are rapidly decaying, the carbon dioxide content of the soil air may exceed 10 percent—300 times the percentage in the outside atmosphere. This carbon dioxide combines with the soil water to produce carbonic acid, which in turn dissolves some of the limestone through which it moves down toward the cave. When the percolating water encounters the air in the cave, which generally has a partial pressure of carbon dioxide much lower than that of soil air, the carbon dioxide leaves the water. It does this for the same reason that it leaves a bottle of a carbonated beverage when the pressure is released by removal of the sealed cap. When the carbon dioxide escapes, a chemical change results, expressed by the following reaction:

$$\underset{\text{calcium bicarbonate solution}}{Ca^{2+} + 2HCO_3^{1-}} \rightarrow \underset{\text{carbon dioxide}}{CO_2\uparrow} + \underset{\text{calcite}}{CaCO_3} + \underset{\text{water}}{H_2O}$$

This process is the reverse of that by which limestone is dissolved to produce caves.

That the loss of carbon dioxide, rather than the evaporation of water, is the chief means by which calcite speleothems are formed is indicated by their chemical composition. Stalactites are almost pure calcite, even though the water from which they form may contain large amounts of other components in solution. Because loss of carbon dioxide results in deposition only of calcium carbonate, the other components are not deposited, as they would be if the water evaporated, and they flow away with the dripping water.

When acidified soil water begins its journey down to the cave, the water immediately begins to dissolve limestone from the walls of the cracks through which it moves. (See the diagram on page 16.) This process continues all the way down to the ceiling of the cave, but the water moves down so fast that when it enters the cave, it is still undersaturated with calcite. We know this to be true because, even though deposition may occur on the outside of speleothems, water trapped inside these same speleothems commonly dissolves interior cavities.

The higher parts of caves, where the water comes directly from the soil zone, have the largest speleothems. The rate at which carbon dioxide is produced in the soil is related to the activity of the microorganisms causing decay; this activity, in turn, depends on the temperature. Because decay is much more rapid when the soil is warm, there is an annual cycle in the carbon dioxide content of cave air. We found, for example, that the carbon dioxide content of a room in Breathing Cave, Virginia, increased from 0.037 on January 1, 1958, to 0.071 percent on May 25, 1958.

A related annual cycle exists with respect to the carbonic acid content of cave water. Acid content is normally recorded in pH units (a measure of the hydrogen ions in solution). Water with a pH value of less than 7.0 is conventionally called acidic, with a value of more than 7.0, alkaline. Water in Tumbling Rock Cave, Alabama, had a pH of 8.2 on January 18, 1958, and of only 5.8 on May 26, 1958. In other words, a much smaller amount of acid was being produced in the winter than in the spring.

There is likewise a correlation between carbon dioxide production and latitude—more carbon dioxide is produced in the thick soil above warm southern caves than in the thin soil above cold northern caves. A possible result of this correlation is that most southern caves contain massive speleothems, whereas northern caves contain only small ones or none at all.

Both dissolution and deposition go on rapidly in warm climates, but which of these processes predominates on a given speleothem at a given time depends chiefly on the rate of loss of carbon dioxide. If the carbon dioxide is not removed rapidly enough, deposition of calcite will cease, and the speleothem may even begin to dissolve. When this happens, the acidic water cuts into the surface and lays bare the raw edges of the growth layers, which then stand in relief like the raised grain of weathered wood.

Stalactites and Related Deposits

CAVE DEPOSITS formed from dripping water are familiar to everyone. The shapes of these structures are determined in part by the shape of the drop and by gravity's effect on it before it falls. A drop of water issuing from a crack in the ceiling of a cave is pulled downward by gravity and hangs from a very small area of attachment. Deposition can occur only in this area. As pressure is reduced on water when it enters a cave chamber, it gives off carbon dioxide and therefore deposits calcite. Because the loss of carbon dioxide begins at the surface of the drop of water, deposition begins where the surface of the drop is in contact with the rock of the ceiling. The calcite is laid down there as a tiny ring. Ring is thus added onto ring to form a hollow cylinder that has the same diameter as the drops of water falling from it. Water continues to move from the crack down through the central hollow of this cylinder to its end, each time adding a small increment of length to the tip of the stalactite. The result is a *tubular stalactite,* often called a "soda straw."

Tubular stalactites are common in caves. Averaging about 5 millimeters in diameter, they may be a meter or more in length, but their walls are only about half a millimeter thick.

The crystal structure of tubular stalactites is simple. They are composed of nearly pure calcite, and each stalactite generally consists of a single crystal. Every time a new layer is added to the end of the tube, the molecules arrange themselves in precise accordance with the pattern laid down during deposition of the previous layer. The tube thus has the same molecular structure as it would if it had been carved from a single enormous calcite crystal. One evidence of this is that where a tubular stalactite has been broken, the breaks follow smooth diagonal cleavage planes, which generally have the same slope wherever fracture has occurred.

In rare cases, a tubular stalactite has a hexagonal cross section. More commonly, the crystal faces of calcite are expressed as minute cusps on the tip of the stalactite. Usually there are three of these cusps, following the trigonal crystal structure of calcite.

Tubular stalactites are so fragile that they sometimes break from their own weight. One occasionally finds a pile of naturally broken stalactites cemented to the floor below a group of slender stalactites hanging from the ceiling. It is likely, however, that when stalactites break from their own weight, the fracture occurs at the junction between two crystals or at some flaw, for laboratory experiments have shown that a perfect specimen will support as much as 5 kilograms—far greater than the weight of any fallen stalactites that have been found. In Soldiers Cave, California, a group of long tubular stalactites is hanging from the ceiling above a bank of soft silt, in which many fallen stalactites are sticking up like stray arrows.

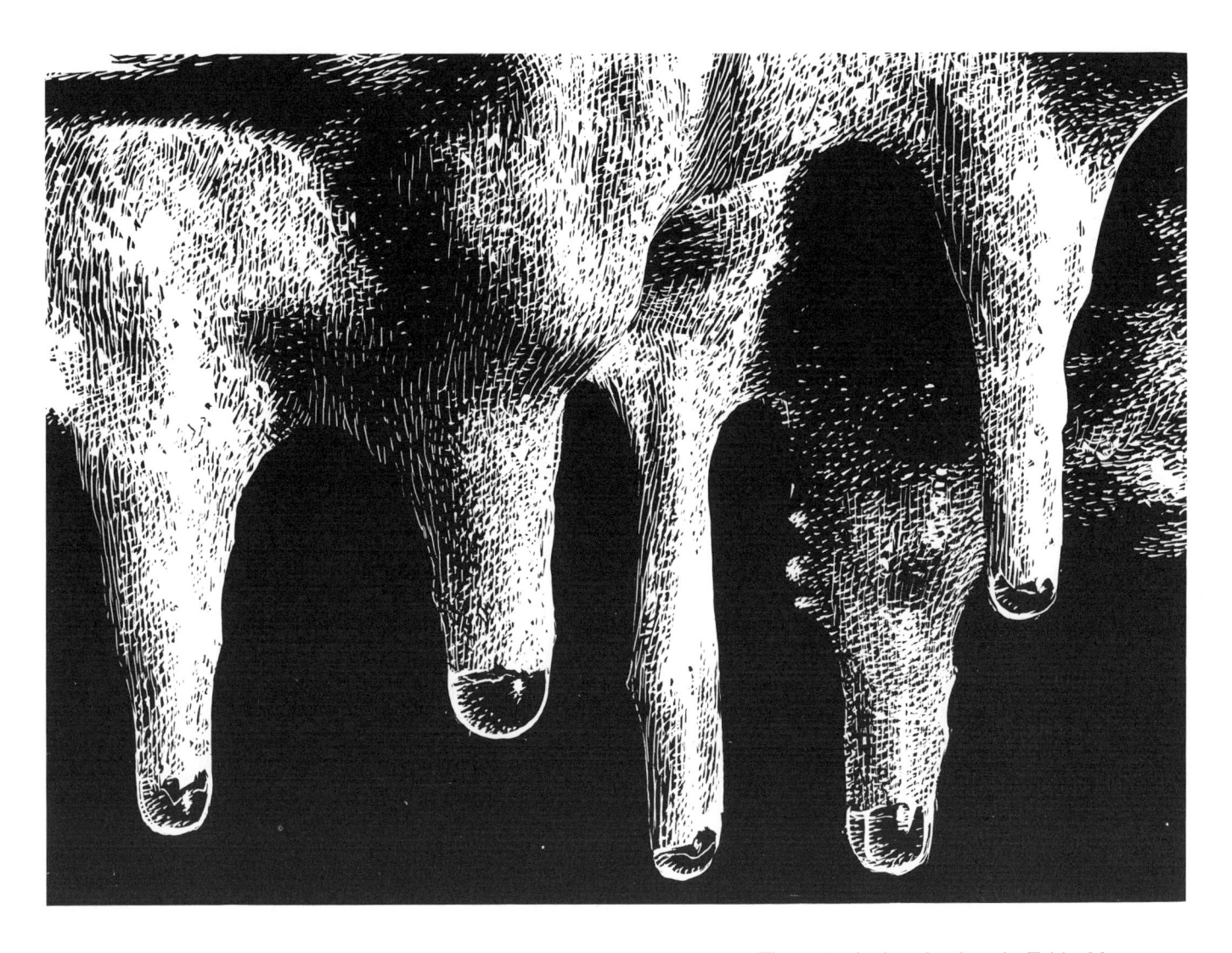

These typical stalactites in Table Mountain Cave, Wyoming, are drawn four times actual size. The blunt tips provide supports for hanging drops of water.

A stalactite begins growing as a small ring of calcite where the surface of a water drop intersects the ceiling of a cave. This ring grows into a tube, which often acquires a tapering shape when water flows down its outer surface.

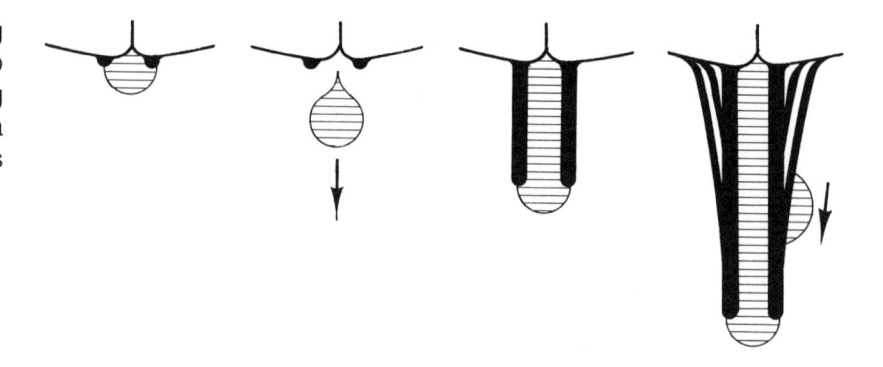

As every visitor to a cave can see, most stalactites are conical, and many are very large. Even huge conical stalactites, however, began as delicate tubular structures. Evidence of this early history can often be found in a broken specimen that shows either a central canal or a zone of clear calcite approximately 5 millimeters in diameter, which represents a former tube that has been filled.

To visualize the process of growth, we must bear in mind that stalactites in caves grow very slowly. Many people have found that stalactites forming on concrete or mortar outdoors may grow several centimeters each year. Stalactite growth in these environments, however, bears little relation to that in caves, because it does not proceed by the same chemical reaction. Although cement and mortar are made from limestone, the same rock in which caves form, the carbon dioxide has been driven off by heating. When water is added to these materials, one product is calcium hydroxide, which is about 100 times as soluble in water as calcite is. A calcium hydroxide solution absorbs carbon dioxide rapidly from the atmosphere to reconstitute calcium carbonate and produce stalactites. This is why stalactites formed by solutions from cement and mortar grow much faster than those in caves. To illustrate, in 1925, a concrete bridge was constructed inside Postojna Cave, Slovenia, and adjacent to it an artificial tunnel was opened. When we visited the cave in 1956, tubular stalactites 45 centimeters long were growing from the bridge, while stalactites of the same age in the tunnel were less than 1 centimeter long.

The best way to learn about the growth rate of cave stalactites is to make repeated measurements. This has been done many times in well-known caves. The data show that the rate of growth is variable, but it is never much over 2 millimeters a year and may average only a little more than a tenth of a millimeter a year.

Number of Years Required for an Increase of 1 Centimeter in Stalactite Length and Stalagmite Height
(From measurements by Martin Kríz, W.B. Dawkins, J.C. Coleman, and W.D. Johnston.)

CAVE	STALACTITE	STALAGMITE
Slouper Cave, Czech Republic	3	33
Inglelborough Cave, England	58	2
New Cave, Ireland	5	—
Grand Caverns, Virginia	—	10

The outer parts of conical stalactites are formed not by dripping water but by flowing water. When water moving across the ceiling begins to flow down the outside of a tubular stalactite, the water deposits calcium carbonate on the surface, most abundantly at the top. It thus not only enlarges the stalactite, but also gives it a conical shape, tapering downward. The deposits on the outside of stalactites are in layers almost parallel to the surface. In cross section, these layers appear as rings made visible by different amounts of impurities in the different layers. These rings are commonly incomplete, and attempts to use them for dating cave deposits have generally been unsuccessful.

The deposits on the outside of a stalactite do not follow the crystal structure of the central tube. Instead, these crystals grow outward everywhere from the surface of the tube. The result is a radial structure made up of small wedge-shaped crystals with their apexes adjacent to the tubular stalactite. As a conical stalactite becomes large, the polygo-

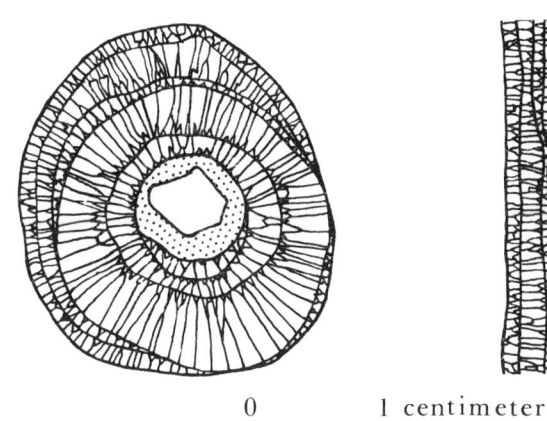

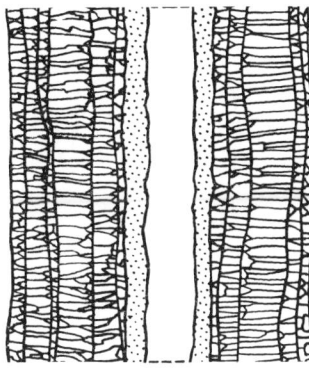

Crystal structure of a stalactite from Cumberland Bone Cave, Maryland. The central tube is dotted. There are many small crystals at the inside of each growth layer, and only a few at the outside of the layer.

Crystal structure of a slice of a stalactite, showing the radiating pattern of the crystals in the outer layers and the vertical orientation in the wall of the central tube.

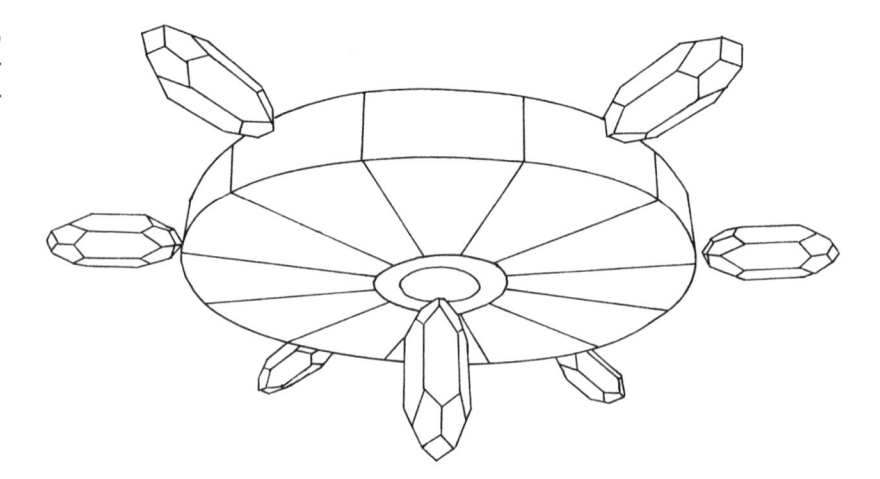

nal bases of these crystals on the outer surface may be as much as 2 centimeters in diameter. They often sparkle beautifully when the light of a lamp is moved across them.

The reason that the crystals in conical stalactites are perpendicular to the surface of growth (other speleothems formed by flowing water have this same structure) is that calcite crystals do not grow at an equal rate in all directions. When allowed to grow freely they form long, pointed crystals called "dogtooth spar." These crystals tend to grow most rapidly in the direction of the long axis. Thus when the crystals are confined, as they are in most speleothems, those whose long axes are nearly perpendicular to the surface tend to grow fastest.

After a dry period, the surface of a speleothem may become coated with a thin layer of impurities, mostly iron oxide minerals and gypsum. When growth resumes, this layer of impurities prevents the newly forming crystals from orienting themselves with the large crystals of the previous layer. As a result, thousands of randomly oriented microcrystals of calcite are formed on the surface. As the deposit begins to grow outward, the crystals that by chance are oriented perpendicular to the surface of growth grow faster and tend to spread over the tops of those less favorably oriented. Those crystals that are parallel with the surface of growth are very soon covered, and the struggle continues among the others until finally only those oriented exactly perpendicular to the surface continue to grow. By this time the crystals are large enough to be seen easily without a microscope; they may grow many centimeters long.

A type of speleothem that owes its shape to a very special mechanism is the *drapery*. Draperies are thin, translucent sheets of calcite that hang down from the ceilings of caves, often 3 meters or more. One of these begins to form when a drop of water flows down an inclined ceiling and leaves behind it a sinuous trail of mineral matter. This drop is followed by another, and the deposit is slowly built downward, layer by layer. During certain periods in the growth of draperies, impurities in the water may leave orange or brown streaks. These impurities give the deposit a banded structure that reminds one of nicely fried bacon.

A drapery just beginning to form is only about as thick as a drop of water. It is made up of long slender crystals of calcite perpendicular to the direction of growth. These crystals are commonly several centimeters long, as thick as the drapery, and about 5 millimeters wide. As each new layer is laid down along the edge of the drapery, the molecules follow the orientation of the adjoining crystals in the previous layer.

In some caves, draperies have serrated edges that have been called fringes, but are perhaps more like the edge of a saw. Each tooth is a crystal.

Stalagmites and Related Deposits

CAVE DEPOSITS formed by flowing and splashing water in the air-filled parts of caves are much more abundant than all other speleothems combined. Many caves contain tons of this material, and some have been nearly filled by it.

Stalagmites are counterparts of stalactites; they rise from the floors of caves. Although their positions are determined by falling water, their shapes are not controlled to any great extent by hanging drops of water. Their form and crystal structure, however, are determined mostly by the same processes that control the shapes of other deposits made up of crystals precipitated from flowing water.

As they fall from stalactites, drops still retain some excess carbon dioxide. The shock of a drop striking the floor and breaking up into droplets or films, with more surface area than the original drop had, causes the gas to be driven off, just as it is when carbonated water is spilled on the floor. This suddenly reduces the solvent power of the water, so that calcite is precipitated to contribute to the building of a stalagmite. A stalagmite grows upward under a stalactite growing downward, and the two may meet to form a *column*.

Stalagmites are usually larger in diameter than the stalactites under which they form, and they generally have rounded tops instead of pointed tips like stalactites. Stalagmites have no central tube. In places, they grow to be more than 15 meters tall and 10 meters in diameter.

When drops fall from great heights, they commonly form flat-topped stalagmites, which are usually made up of a series of slightly offset plates, about a centimeter thick and averaging about 20 centimeters in diameter. A factor that may be significant here is that flat-topped stalagmites are generally found in vertical shafts that communicate almost directly with the surface. For this reason, water in vertical shafts usually has dissolved less limestone on the way down to the cave than most cave water has. Hence, the water in the shafts is less saturated than most cave water. It still has a slight dissolving capacity as it falls from the ceiling. This is dissipated almost immediately after it strikes the stalagmite, owing to the loss of carbon dioxide, but the brief period of contact with an undersaturated solution causes the top of the stalagmite to become flat or even cup-shaped, rather than being rounded like normal stalagmites.

The crystal structure of stalagmites is similar to that of the outer part of stalactites. Rarely, if the dripping solutions have few impurities, a stalagmite forms as a single crystal, but ordinarily the structure is radial, consisting of many crystals perpendicular to the surface of growth. Wedge-shaped and pyramid-shaped crystals broaden outward from the interior.

The rate of growth of stalagmites seems to be about the same as that of stalactites. Stalagmites grow in height at a maximum rate of a little more than 2 millimeters a year and probably average a few tenths of a millimeter a year.

WHEN WATER flows down the walls of a cave, sheets of calcite called *flowstone* are laid down. The crystals in flowstone are oriented perpendicular to the surface of deposition, and sometimes very large individual crystals grow on the outer surface of a deposit. The surface is smooth, though, and the presence of the crystals can be detected only by the way in which they sparkle.

In some caves, flowstone forms over gravel or silt beds. In places, the gravel or silt is later washed out from under the flowstone, producing unsupported hanging deposits called *canopies*, which often have impressions of mud cracks on their undersurfaces. Beautiful hanging deposits are formed when water continues to run over the flowstone, causing a deposit of stalactites from the lip of the canopy.

Cave pearls in Carlsbad Cavern, New Mexico, are shown here four times actual size. Each pearl was formed by deposition around a nucleus while it was agitated by dripping water.

Rimstone dams are curious, although fairly common deposits in caves. Located on cave floors, these dams are walls that impound small pools of water or in some cases dam cave streams. The dams usually occur in a series of curved steps that impound crescent-shaped pools; water flows over them from one pool to another. Rimstone dams range widely in size; some are only a few millimeters high, whereas others are more than a meter.

One hypothesis for the origin of rimstone dams is that as water flows over a rapids it is slightly agitated, causing carbon dioxide to be given off. This results in deposition of calcium carbonate that starts the lip of a dam. Because more water flows over low parts of the dam than elsewhere, more calcite is deposited there, and the top of the dam therefore remains nearly level. Water impounded in the pool becomes supersaturated with respect to calcite, because carbon dioxide continually diffuses from the water surface, but deposition is delayed until it is nucleated by calcite crystals on the dam.

One of the most interesting types of speleothems is the *cave pearl*. The famous French speleologist Norbert Casteret has called cave pearls the rarest of all cave deposits. One of the reasons these small spherical bodies are not often seen is that they are loose, and visitors can carry them away from the cave.

Cave pearls range from grains of pinhead size to irregular bodies as much as 15 centimeters in diameter. Most of the smaller ones are nearly spherical, whereas the large ones tend to be irregular. Yet a nearly perfect sphere found in Carpenter Cave, Virginia, is almost 8 centimeters in diameter and weighs 1 kilogram. Cave pearls and similar objects formed in the ocean are called *pisolites* (from the Latin word for peas) if they are more than 2 millimeters in diameter, and *oolites* (from the Greek word for eggs) if they are smaller.

Cave pearls generally have a nucleus, which may be a grain of sand or a fragment of another speleothem. Surrounding this nucleus, concentric layers, usually of calcite, are laid down in much the same fashion as layers are deposited on an oyster pearl. As in other material formed by flowing and splashing water, and as in oyster pearls, the crystals are perpendicular to the surface of growth.

Many cave pearls are formed in shallow pools below dripping water. Constant agitation and rotation seems to be necessary for the formation of large spherical pearls. Many large pearls, however, are so irregular in form that they cannot have been continuously rotated, and rotation is not essential to keep a pearl from attaching itself to the floor. Some smaller pearls are nearly cubical. In places, large numbers of these nestle together, and although they cannot rotate or even move much with respect to one another, the pearls maintain their individuality and seem to grow equally on all sides. William Emmons described a steel machine nut, which was incapable of rotation, coated with calcite in a mine; as much calcite was deposited on the bottom surface as on the top, yet the nut was not cemented to the floor.

Deposits Formed by Seeping Water

AMONG THE MOST INTERESTING OBJECTS found in caves are those formed by seeping water. These curiously shaped speleothems, some of them extremely delicate and beautiful, project from the walls in a way that seems almost to defy gravity.

Among the best known types of objects formed by seeping water are *helictites*—small twisted structures usually consisting of calcite. They are generally several centimeters or more long and about 5 millimeters in diameter. Because they project from ceilings, walls, and floors of caves at many angles, some investigators have called them "eccentric stalactites." Helictites have excited the imagination of speleologists for many years. Perhaps more hypotheses of origin have been proposed for them than for any other class of speleothem. Some of the most improbable ideas that have been presented are that these curious objects result from deposition on spider webs or fungi, that they condense directly from a vapor that contains lime, or that they curve because of electrical energy.

The first bit of evidence contributing to a better understanding of the origin of helictites and their curious behavior was presented in 1894 by George Merrill, who observed that each helictite has a narrow central canal extending along its axis. This canal, which is about 0.2 millimeter in diameter, conducts water under hydrostatic pressure from a minute opening in the wall of the cave through the helictite to its tip. The flow is so slow that a drop of water does not form, and gravity is not given an opportunity to affect the shape. Most of the deposition takes place right around the hole in the tip of the helictite. Each new layer is shaped like a tiny cone, and each cone consists of a single crystal. Their crystal form causes the cones to be slightly distorted, so that a new cone never fits perfectly on the one it covers. Hence each cone is systematically tilted, and the whole helictite assumes a twisted or spiral shape.

Calcite helictites that are free from impurities are often nearly transparent. Their surfaces are fairly smooth, and their trigonal crystal struc-

Helictites are deposited from water that moves in a nearly microscopic central canal from the point of attachment to the tip. This calcite helictite in Soldiers Cave, California, is shown about four times actual size.

Aragonite helictites form in the same way as calcite helictites, by seepage through a central canal and deposition on the tip. These examples in Northumberland Cave, Nevada, are shown about twice actual size.

ture is reflected in a triangular cross section near the tip and a hexagonal cross section at positions along the body.

Some helictites are made of the mineral aragonite, composed like calcite of calcium carbonate, but having a different crystal structure. The surfaces of aragonite helictites are rough, being made up of many small pointed crystals radiating outward from the central canal. Some aragonite helictites have a beaded shape, with about four beads to a centimeter. If each bead represents an annual cycle, the helictites grow rather rapidly. There are beaded aragonite helictites in Cave of the Winds, Colorado, and a few in the New Mexico Room of Carlsbad Cavern.

Calcite helictites occur in many caves throughout the world, but aragonite helictites are less common. The best known aragonite helictites in the United States are in Skyline Caverns, Virginia. These are somewhat complicated, however, by having a coating of calcite on their tips.

Another speleothem related in origin to helictites is the *shield*, also formed by seeping water. Shields are semicircular sheets averaging about 3 centimeters in thickness and a meter in diameter. Each shield is attached along its straightest edge to the ceiling, wall, or floor of a cave, and they project outward at various angles into the chamber. The rims of shields are commonly ornamented with stalactites and draperies; helictites often form on their upper surfaces.

The now-accepted concept of the origin of shields was proposed in 1950 by the Czech speleologist Josef Kunsky.

Each shield consists of two parallel plates separated by a planar fracture. This fracture is always an extension of a joint or parting in the limestone bedrock. The arrangement of the growth bands in broken specimens shows that the most rapid growth occurs where the internal crack intersects the edge of the shield. As in helictites, water under pressure and containing dissolved calcium carbonate moves along the crack between the two plates out to the rim of the shield. Here deposition occurs around the edges of the plates, and the shield slowly increases in diameter. Because hydrostatic pressure controls the flow of water, shields may form at any angle on the walls, or may even grow upward from the floor.

Shields are relatively rare. The best-known examples in the United States are in Grand Caverns, Virginia, and Lehman Caves, Nevada.

One of the most common of all cave deposits is *cave coral*. These small, knobby clusters are found in nearly every cave. The generally smooth individual knobs are about 5-10 millimeters in diameter. Some knobs are on stems that branch, and they usually stand several centi-

Vertical cross section of a specimen of cave coral from Soldiers Cave, California, enlarged 5 times. The growth layers are nearly parallel with the rounded upper surfaces, and the crystals are perpendicular to those surfaces. The water is thought to have seeped upward between the crystals to deposit its load of calcite on the upper surfaces.

meters from the wall, although some may extend outward 15 centimeters or more. In cross section, the knobs in cave coral exhibit a concentric banded structure. The radiating crystals are perpendicular to the growth bands.

Cave coral often occurs in places where it could not have received water from dripping or splashing. Despite this seeming lack of water source, some of these deposits have wet surfaces and seem to be growing actively. Cave coral most commonly occurs along cracks in a wall or on porous deposits of cave silt. In some places, tubular stalactites hang from the outward terminations of the cave-coral knobs. These facts suggest that cave coral is formed by seeping water, but as no central canals have been observed thus far, the water apparently seeps out from between the crystals. To demonstrate that the water can move through cave coral, one can place the base of a specimen in a copper sulfate solution; within a few hours blue crystals of copper sulfate will begin to form on the tops of the knobs.

A kind of speleothem that may be related to cave coral is the *spherical stalactite*. This does not have the more usual conical shape, but is bulbous. Some examples are nearly spherical, and others are made up of two or more connected spheres. Well-developed specimens in the United States are in Massanutten Caverns, Virginia, where spherical stalactites average 15 centimeters in diameter. Internally, they have concentric growth bands and a radial crystal structure. They are porous and their cavities seem to have been dissolved after the stalactites were formed. Evidently, water seeps outward from the interior and deposits calcite on the surface.

In another group of speleothems formed by seeping water, the material is not deposited at the free end, as in helictites, but at the attached end. *Cave flowers* are the most common example of this type. Most cave flowers are composed of gypsum ($CaSO_4 \cdot 2H_2O$). They sprout from the walls of dry cave rooms in a curving fashion, much like helictites. The principal structural difference between cave flowers and helictites is that the flowers have longitudinal striations on their surfaces and possess no central canals. Cave flowers average about 5 centimeters in length. Several of them commonly diverge from a center, like petals. Rarely, as in Mammoth Cave, Kentucky, they may project more than 30 centimeters from the wall.

The crystals of cave flowers form in minute pores in the rock and then are forced out into the cave chamber by more crystals forming behind them. As the mass of crystals moves out, much like toothpaste squeezed from a tube, it usually grows more rapidly on one side than on the other. This is what gives cave flowers their delicate curves.

Some caves contain extruded filaments of gypsum so slender and threadlike that they wave like spider webs with each breath of air. Massive bunches of these filaments are called *cave cotton*. This leads us to one of the most bizarre of all the strange mineral deposits found in caves. On a wall of Silent River Cave, Arizona, a compact, ropelike cluster of these gypsum filaments extends down to the floor of the cave. The "cave rope" is about 2 centimeters in diameter, and where it reaches the floor it has formed a neat counterclockwise coil about 15 centimeters in diameter!

Cave blisters are hollow hemispherical deposits that form on the walls of caves. The diameter of the blisters averages about 5 centimeters but may be as much as 30 centimeters. Each consists of a thin shell, usually of gypsum, enclosing a powdery mixture of calcite and other minerals. It is thought that cave blisters form like cave flowers, except that the crystals are extruded from an initial bud in a circle of ever-increasing circumference.

Crystals of gypsum also form within some deposits of silt that partly fill caves. These crystals are generally irregular, but some are arrow-

Cave flowers are extruded from the walls of a cave much as toothpaste is pushed from a tube. The mineral composing a flower is deposited at the attached end, in pores in the wall rock. These gypsum cave flowers in Cumberland Caverns, Tennessee, are shown about half actual size.

shaped, or shaped like the silhouette of a fir tree. In some caves, fairly regular spinelike twin crystals are extruded from the silt in the fashion of frost crystals. All these gypsum crystals are formed by seeping water moving through the silt.

Deposits Formed in Standing Water

ALTHOUGH SPELEOTHEMS formed on the surfaces of cave pools make up only a very small part of all cave deposits, they include some of the more interesting species. One of these is the *cave raft*, a thin floating film, usually of calcite. Cave rafts, which are usually not much more than a tenth of a millimeter thick, lie supported by surface tension on the surfaces of pools. If the surface of the water is disturbed, the cave rafts slowly sink to the bottom of the pool. In Goshute Cave, Nevada, is a deposit, more than 30 centimeters thick, made up of fragments of these rafts in an area once covered by a cave pool.

At the margins of some cave pools, deposits attached to the wall grow out over the water. Some cave pools are completely covered by these shelves, which are similar in origin to cave rafts.

Another type of speleothem formed on the surface of cave pools is the *cave bubble*, which Gordon Warwick was the first to recognize underground in a limestone mine in England. Similar bubbles sometimes occur also on pools associated with surface hot springs.

Cave bubbles, never more than 5 millimeters in diameter, have very thin walls. They are formed at the surface of the water by calcite crystallizing around bubbles. They usually float on the surfaces of pools behind rimstone dams; as deposition continues, the bubbles may become too heavy to float, and then they sink to the bottom of the pool.

Calcite cave bubbles that we studied in Goshute Cave, Nevada, are not precisely spherical, but are cup shaped, with slightly domed covers. The walls of the lower part of the cup are about 0.2 millimeter thick, but the cover is much thinner. Each of the specimens had a hole in the center of the top. The specimens were dry when we examined them; at the time they were forming, this hole was probably covered by a lens of water.

Another environment in which speleothems form is underwater in cave pools. Here one finds the best-developed cave-mineral crystals, especially of calcite. Sometimes after a cave has been formed, it may be filled with water, and calcite crystals may grow from the walls into the water. In time, the entire cave is turned into an enormous ge-

Large calcite crystals may form on the walls of caves that remain partly inundated after air circulation has begun to remove excess carbon dioxide. The crystals of "dogtooth spar" shown here are in Sitting Bull Cave, South Dakota.

ode, lined with large crystals projecting inward. A good example of this type of cave in the United States is Sitting Bull Cave, South Dakota. Its walls are covered with pyramid-shaped crystals of calcite as much as 15 centimeters long. In several of the nearby caves of the Black Hills region, as many as four cycles of inundation and crystal formation can be recognized.

Although calcite is the most common mineral in underwater deposits in caves, other minerals also sometimes grow under water. Naica Cave, Mexico, exhibits beautiful transparent bladed crystals of gypsum that commonly contain water-filled cavities. These crystals average about 2 centimeters in diameter; some are nearly a meter long.

Ancient Climate Recorded by Speleothems

AS WE DISCUSS in Chapter 3 on characteristics of the underground atmosphere, surface temperature variations are suppressed in caves. In the upper levels of a cave where most speleothems form, the cave temperature represents the average surface temperature over the preceding few years. Therefore, long-term surface climatic change can be recorded by speleothems during their formation, and research in the past 20 years has shown that speleothems are among the best indicators of the ancient climate of land areas.

To study the climate of the past, stalagmites are dated by a radiometric method, usually by the thorium-230 method. Natural water contains a minute amount of uranium that is held in solution as a complex ion, the uranyl ion (UO_2^{2+}). Because UO_2^{2+} shares with Ca^{2+} a similar size and the same combining power (their shared 2+ valence), some of the calcium in the calcite of speleothems is replaced by a small quantity of uranium.

The various isotopes of uranium decay radioactively through a series of radioactive daughter elements. One step in the decay chain is the decay of uranium-234 into thorium-230. Inasmuch as cave calcite initially contains practically no thorium (because thorium in solution quickly adheres to soil particles and hence does not reach the cave), any thorium-230 that is found in a speleothem is inferred to have formed in place from uranium-234 after the speleothem was deposited.

Despite the very small quantities of uranium and thorium in stalagmites, speleologists can detect their isotopes very precisely by measuring the amount of radioactivity and its energy, which is diagnostic for each isotope. This method permits accurate dating by the thorium-

230 method up to an age of about 300,000 years. Beyond that, the buildup of thorium-230 ceases, because greater amounts are destroyed by their own radioactivity as fast as they are produced from uranium-234.

Thousands of speleothem samples in various climatic zones have now been dated, thus making it feasible to study paleoclimates. Closely spaced sequential samples along the vertical axes of stalagmites show periods of growth alternating with periods of nongrowth, the latter usually representing droughts. As the body of data has grown larger, we have been able to separate local effects from regional ones, and to map the changes in precipitation patterns during the last several glacial stages of the Pleistocene Epoch.

In addition to showing past changes in precipitation, the calcite of speleothems also records past changes in temperature by changes in the ratio of its oxygen isotopes. Oxygen, chiefly composed of oxygen-16, contains about 0.2 percent of oxygen-18. In calcite, this percentage varies with the temperature of formation, because oxygen-18 concentrates in calcite relative to water, and the concentration becomes less with increasing cave temperature.

Oxygen-18 values in cave calcite, in addition to being related to the temperature of calcite formation, are also related to the oxygen-18 content of the original depositing water, which varies with other aspects of the climate besides temperature. To obtain the most accurate temperature values, speleologists analyze small quantities of the water that is almost always trapped and preserved in stalagmites in fluid inclusions—sausage-shaped cavities 10–50 micrometers long. This trapped water has lost its original oxygen isotope record by exchange with the oxygen of the enclosing calcite, but hydrogen isotope analysis can be used from a well-known relationship between hydrogen and oxygen isotope ratios to estimate its original oxygen-18 content. When compared with the oxygen-18 content of samples of the enclosing calcite, these values give a precise measure of the ancient temperature of the cave and of the surface above it.

Research remains very active in this field. In recent results, speleothems indicate that the average surface temperature in midlatitude cave regions reached a peak 3°C above the present about 8,000 years ago, that it was as much as 10°C colder than at present from 15,000 to 80,000 years ago, warmer than now from 80,000 to 120,000 years ago, colder from 120,000 to 170,000 years ago, and colder for an undetermined period before that. Further studies should give us a detailed picture of climatic change in the world during the past 300,000 years.

Cave Minerals

A TOTAL OF 175 DIFFERENT MINERALS, a few of which have not been found anywhere else, occur in limestone caves. Most cave deposits, however, are composed of only a few mineral species. Calcite greatly predominates, followed in approximate order of abundance by gypsum, ice, aragonite, the iron mineral goethite, and the manganese mineral birnessite. Other minerals make up only a very small percentage of all speleothems, but in some caves one or more of them are present in sufficient quantity to constitute a commercial deposit.

More than 40 of the known cave minerals contain phosphorus derived from bat droppings. Such deposits of bat droppings—called *guano*—are gathered from caves in some places for use as fertilizer. The cave bats that produce guano deposits eat insects, and their droppings consist chiefly of the indigestible horny shells of these insects. The shells are composed of chitin, a substance rich in phosphorus. Another contributor to the chemical instability of the guano, and to its value as a fertilizer, is the nitrogen-rich urine of the bats. Fresh bat guano contains about 12 percent nitrogen and 4 percent phosphorus.

Cave materials of several types may interact to produce cave minerals. For example, solutions from zinc ore, which occurs in limestone in places, may react with organic cave deposits to produce the zinc and phosphate minerals shown here.

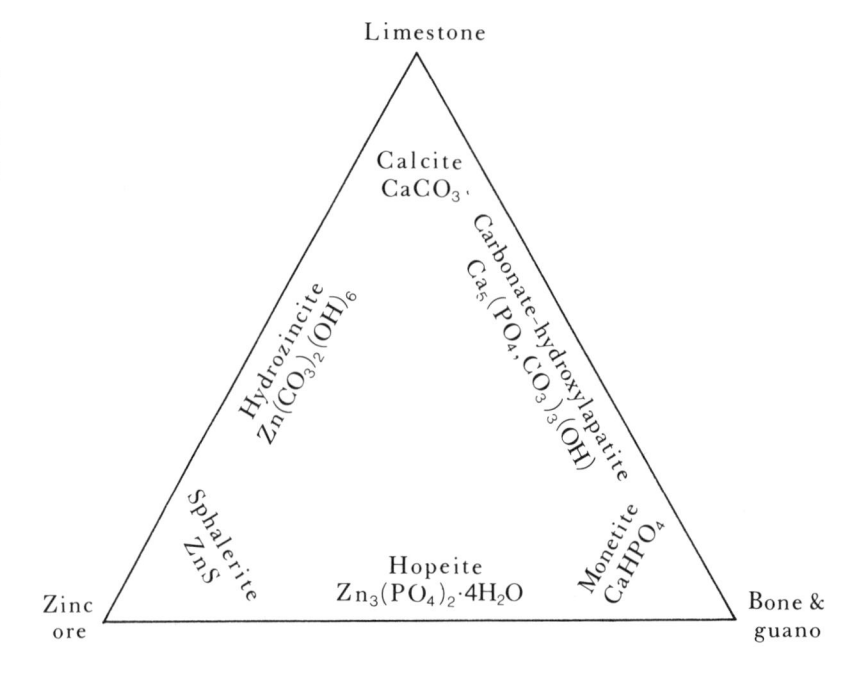

Limestone

Calcite $CaCO_3$

Hydrozincite $Zn(CO_3)_2(OH)_6$

Carbonate-hydroxylapatite $Ca_5(PO_4,CO_3)_3(OH)$

Sphalerite ZnS

Hopeite $Zn_3(PO_4)_2 \cdot 4H_2O$

Monetite $CaHPO_4$

Zinc ore

Bone & guano

One of the earliest decomposition products of the bat guano is ammonia, whose pungent odor is readily detected in caves containing large numbers of bats. The decomposition and leaching of the guano produces a series of mineral products. Those that form first are rich in nitrogen and those that form last are rich in phosphorus. The final product in this series of minerals is usually carbonate-hydroxylapatite, a fairly stable phosphate mineral formed by reaction between solutions from the guano and the calcite in the wall rock of the cave.

In some cave rooms, reddish brown phosphatic speleothems occur where a source for the phosphate is not clearly evident. For example, the ceiling of the Red Room of Muierii Cave, Romania, is covered by stalactites and draperies of carbonate-hydroxylapatite. Source materials do not occur in that room, but accumulations of guano are known elsewhere in the cave. In a recent study of the cave, G. Diaconu and A. Medesan found by careful mapping that the Red Room lies 20 meters directly below the Alter Room, which contains one of the largest deposits of guano in the cave, and which also contains considerable bone debris. Phosphate leached from these materials has percolated down through fractures in the limestone and reacted with it to produce the speleothems in the Red Room below.

AS MENTIONED ABOVE, one mineral that occurs fairly abundantly in caves is aragonite. This mineral has the same composition as calcite ($CaCO_3$), but its atoms are differently arranged, and the two minerals consequently differ in their crystal form and in many other properties. One difference is that aragonite is about 10 percent more soluble than calcite in laboratory experiments performed at cave temperatures. It follows, then, that cave water slightly supersaturated with respect to calcite could deposit calcite but could only dissolve aragonite.

Aragonite would never form in caves if cave water were not often so highly supersaturated with respect to calcite that it can no longer dissolve aragonite. This very highly supersaturated water could deposit either aragonite or calcite, or both, but what such water does deposit in caves is usually aragonite.

Aragonite crystals nucleate faster than calcite crystals in water greatly supersaturated with respect to both minerals. This by itself would lead to a mixture of the two minerals, a situation that is common in laboratory experiments where impurities are excluded, but uncommon in caves. Most cave water contains 10-100 milligrams per liter of magnesium in solution. Not much magnesium enters the aragonite crystal structure, but it readily enters the calcite structure, because the carbonate mineral of magnesium—magnesite—belongs to the same crystal system. Magnesium increases the solubility of calcite, and along with other impurities, it further slows calcite nucleation, especially in warm water. Because magnesium and other calcite inhibitors are generally present in cave water, and because aragonite nucleating agents are present also, highly supersaturated water in warm caves usually deposits aragonite.

In the New Mexico Room of Carlsbad Cavern, an aragonite stalactite is forming above a calcite stalagmite. The stalactite is evidently growing from a solution that is greatly supersaturated as a result of the sudden loss of carbon dioxide gas that occurs when water enters the cave. The calcite of the stalagmite below is being formed from the residual solution, but the deposition of aragonite on the stalactite has so reduced the level of supersaturation that the solution now deposits calcite rather than aragonite.

The caves with the highest levels of supersaturation are warm southern caves, because bacterial production of carbon dioxide in the overlying soil is very rapid. Aragonite, relatively common in southern caves, is rare in cold northern caves and in caves at high altitudes.

Many caves of intermediate temperature in the United States contain deposits of fossil aragonite—aragonite formed in the past but now being covered by calcite. Because caves in which aragonite is now forming tend to be warm caves, it has been inferred that fossil aragonite records a warmer period in the history of the caves that contain it.

Secondary Minerals in Limestone Caves

Mineral, Formula	Typical Locality	Reference
Allophane, $Al_2SiO_5 \cdot nH_2O$	Mbobomkulu Cave, South Africa	Martini, 1980
Aluminite, $Al_2(SO_4)(OH)_4 \cdot 7H_2O$	Mbobomkulu Cave, South Africa	Martini, 1980
Alunite, $KAl_3(SO_4)_2(OH)_6$	Mbobomkulu Cave, South Africa	Martini, 1980
Anglesite, $PbSO_4$	Ahumada Cave, Mexico	Rickard, 1924
Anhydrite, $CaSO_4$	Diana Cave, Romania	Diaconu, 1974
Aphthitalite, $(K,Na)_3Na(SO_4)_2$	Murra-el-elevyn Cave, Australia	Bridge, 1975
Aragonite, $CaCO_3$	Skyline Caverns, Virginia	Henderson, 1949
Arcanite, K_2SO_4	Timbavati Cave, South Africa	Martini, 1984
Archerite, $(K,NH_4)H_2PO_4$	Petrogale Cave, Australia	Bridge, 1977
Ardealite, $Ca_2(SO_4)(HPO_4) \cdot 4H_2O$	Csoklovina Cave, Romania	Schadler, 1932
Arnhemite, $K_2MgP_2O_7 \cdot nH_2O$	Arnhem Cave, Namibia	Martini, 1993
Arseniosiderite, $Ca_2Fe_3(AsO_4)_3O_2 \cdot 3H_2O$	Tyuya-muyun Cave, Russia	Smolianinova, 1970
Atacamite, $Cu_2Cl(OH)_3$	Jingemia Cave, Australia	Bridge and others, 1978
Aurichalcite, $(Zn,Cu)_5(CO_3)_2(OH)_6$	Blanchard Cave, New Mexico	Hill, 1976
Azurite, $Cu_3(CO_3)_2(OH)_2$	Copper Queen Cave, Arizona	Ransome, 1904
Barite, $BaSO_4$	Madoc Cave, Canada	Walker, 1919
Basaluminite, $Al_4(SO_4)(OH)_{10} \cdot 5H_2O$	Mbobomkulu Cave, South Africa	Martini, 1993
Bassanite, $2CaSO_4 \cdot H_2O$	Flower Cave, Texas	Hill, 1979
Beudantite, $PbFe_3(AsO_4)(SO_4)(OH)_6$	Island Ford Cave, Virginia	Dietrich, 1960
Biphosphammite, $NH_4H_2PO_4$	Murra-el-elevyn Cave, Australia	Bridge, 1977
Birnessite, $CaMn_7O_{14} \cdot 3H_2O$	Weber Cave, Iowa	Moore, 1981
Bloedite, $Na_2Mg(SO_4)_2 \cdot 4H_2O$	Lee Cave, Kentucky	White, 1971
Bobierrite, $Mg_3(PO_4)_2 \cdot 8H_2O$	Arun Aas Cave, Namibia	Martini, 1993
Boehmite, $AlO(OH)$	Dachstein-Mammut Cave, Austria	Seeman, 1981
Boussingaultite, $(NH_4)_2Mg(SO_4)_2 \cdot 6H_2O$	Gwihaba Cave, Botswana	Martini, in press
Brazilianite, $NaAl_3(PO_4)_2(OH)_4$	Wondergat, South Africa	Cairncross and Dixon, 1995
Brochantite, $Cu_4(SO_4)(OH)_6$	Blanchard Cave, New Mexico	Hill, 1976
Bromargyrite, $AgBr$	Bisbee Cave, Arizona	Graeme, 1981
Brushite, $CaHPO_4 \cdot 2H_2O$	Pig Hole Cave, Virginia	Murray and Dietrich, 1958
Calciovolborthite, $CaCu(VO_4)(OH)$	Tyuya-muyun Cave, Russia	Smolianinova, 1970
Calcite, $CaCO_3$	Grand Caverns, Virginia	Merrill, 1894
Carbonate-fluorapatite, $Ca_5(PO_4,CO_3)_3F$	Poorfarm Cave, West Virginia	Davies, 1958
Carbonate-hydroxylapatite, $Ca_5(PO_4,CO_3)_3(OH)$	El Chapote Cave, Mexico	Pérez and Wiggen, 1953
Celestite, $SrSO_4$	Cumberland Caverns, Tennessee	White, 1958

Cerussite, $PbCO_3$	Herman Smith Cave, Illinois	Bradbury, 1959
Chalcanthite, $CuSO_4 \cdot 5H_2O$	Iron Canyon Cave, Nevada	Roberts and Arnold, 1956
Chalchophanite, $(Zn,Fe,Mn)Mn_3O_7 \cdot 3H_2O$	Bisbee Cave, Arizona	Graeme, 1981
Chalcoalumite, $CuAl_4(SO_4)(OH)_{12} \cdot 3H_2O$	Mbobomkulu Cave, South Africa	Martini, 1980
Chrysocolla, $(Cu,Al)_2H_2Si_2O_5(OH)_4 \cdot nH_2O$	Tyuya-muyun Cave, Russia	Smolianinova, 1970
Cinnabar, HgS	Guadakskoy Cave, Russia	Lazarev and Philenko, 1976
Clairite, $(NH_4)_2(Fe,Mn)_3(SO_4)_4(OH)_3 \cdot 3H_2O$	Lone Creek Fall Cave, South Africa	Martini, 1980
Collinsite, $Ca_2(Mg, Fe)(PO_4)_2 \cdot 2H_2O$	Blue Lagoon Cave, South Africa	Martini, in press
Conichalcite, $CaCu(AsO_4)(OH)$	Bisbee Cave, Arizona	Graeme, 1981
Crandallite, $CaAl_3(PO_4)_2(OH)_5 \cdot H_2O$	Pájaros Cave, Puerto Rico	Kaye, 1959
Cristobalite, SiO_2	Soldiers Cave, California	Moore, 1951
Cuprite, Cu_2O	Bisbee Cave, Arizona	Graeme, 1981
Cyanotrichite, $Cu_4Al_2(SO_4)(OH)_{12} \cdot 2H_2O$	Blanchard Cave, New Mexico	Hill, 1976
Darapskite, $Na_3(SO_4)(NO_3) \cdot H_2O$	Flower Cave, Texas	Hill and Ewing, 1977
Descloizite, $PbZn(VO_4)(OH)$	Tyuya-muyun Cave, Russia	Smolianova, 1970
Devilline, $CaCu_4(SO_4)_2(OH)_6 \cdot 3H_2O$	Monte Rosso Cave, Italy	Chiesi and Forti, 1985
Diadochite, $Fe_2(PO_4)(SO_4)(OH) \cdot 5H_2O$	Feen Cave, Germany	Becker, 1925
Dolomite, $CaMg(CO_3)_2$	Lehman Caves, Nevada	Moore, 1961
Endellite, $Al_2Si_2O_5(OH)_4 \cdot 2H_2O$	Carlsbad Cavern, New Mexico	Davies and Moore, 1957
Epsomite, $MgSO_4 \cdot 7H_2O$	Wyandotte Caverns, Indiana	Blatchley, 1897
Evansite, $Al_3(PO_4)(OH)_6 \cdot 6H_2O$	Westdriefontein Cave, South Africa	Martini, 1978
Ferghanite, $U_3(VO_4)_2 \cdot 6H_2O$	Tyuya-muyun Cave, Russia	Chirvinsky, 1929
Fluorapatite, $Ca_5(PO_4)_3F$	Slaughter Canyon Cave, New Mexico	Hill, 1976
Fluorite, CaF_2	Spirit Mountain Cave, Wyoming	Hixson, 1962
Francoanellite, $H_6K_3Al_5(PO_4)_8 \cdot 13H_2O$	Castellana Caves, Italy	Balenzano and others, 1976
Galena, PbS	Herman Smith Cave, Illinois	Bradbury, 1959
Gibbsite, $Al(OH)_3$	Dachstein-Mammut Cave, Austria	Seeman, 1981
Glushinskite, $Mg(C_2O_4) \cdot 2H_2O$	Temple of Doom Cave, Namibia	Martini, in press
Goethite, $FeO(OH)$	Paxton Cave, Virginia	Hill, 1976
Guanine, $C_5H_3(NH_2)N_2O$	Dingo Donga Cave, Australia	Bridge, 1974
Gwihabaite, $(NH_4,K)NO_3$	Gwihaba Cave, Botswana	Martini, in press
Gypsum, $CaSO_4 \cdot 2H_2O$	Mammoth Cave, Kentucky	Weller, 1927
Halite, $NaCl$	Snake Creek Cave, Nevada	Rogers, 1975
Hannayite, $(NH_4)_2Mg_3H_4(PO_4)_4 \cdot 8H_2O$	Niah Cave, Malaysia	Bridge and Robinson, 1983
Hematite, Fe_2O_3	Wind Cave, South Dakota	White and Deike, 1962
Hemimorphite, $Zn_4Si_2O_7(OH)_2 \cdot H_2O$	Broken Hill Cave, Zambia	Spencer, 1908

Mineral	Cave	Reference
Hewettite, $CaV_6O_{16} \cdot 9H_2O$	Tyua-muyun Cave, Russia	Smolianinova, 1970
Hexahydrite, $MgSO_4 \cdot 6H_2O$	Lee Cave, Kentucky	White, 1971
Hopeite, $Zn_3(PO_4) \cdot 4H_2O$	Broken Hill Cave, Zambia	Spencer, 1908
Huntite, $CaMg_3(CO_3)_4$	Titus Canyon Cave, California	Moore, 1961
Hydromagnesite, $Mg_5(CO_3)_4(OH)_2 \cdot 4H_2O$	Carlsbad Cavern, New Mexico	Davies and Moore, 1957
Hydrombobomkulite, $NiAl_4(NO_3)_2(OH)_{12} \cdot 14H_2O$	Mbobomkulu Cave, South Africa	Martini, 1980
Hydroxylapatite, $Ca_5(PO_4)_3(OH)$	Negra Cave, Puerto Rico	Kaye, 1959
Hydrozincite, $Zn_5(CO_3)_2(OH)_6$	Island Ford Cave, Virginia	Dietrich, 1960
Ice, H_2O	Fossil Mountain Cave, Wyo.	Halliday, 1954
Jarosite, $KFe_3(SO_4)_2(OH)_6$	Tintic Cave, Utah	Stringham, 1946
Kieserite, $MgSO_4 \cdot H_2O$	Tanadival Serrata Cave, Italy	Bernasconi, 1962
Lecontite, $(NH_4,K)Na(SO_4) \cdot 2H_2O$	Piedras Cave, Honduras	Taylor, 1858
Lepidocrocite, $FeO(OH)$	Dachstein-Mammut Cave, Austria	Seeman, 1981
Leucophosphite, $KFe_2(PO_4)_2(OH) \cdot 3H_2O$	Westdriefontein Cave, South Africa	Martini, 1978
Lonecreekite, $(NH_4)Fe(SO_4)_2 \cdot 12H_2O$	Lon Creek Fall Cave, South Africa	Martini, 1983
Maghemite, Fe_2O_3	Dachstein-Mammut Cave, Austria	Seeman, 1981
Magnesite, $MgCO_3$	Titus Canyon Cave, California	Moore, 1961
Magnetite, Fe_3O_4	Dachstein-Mammut Cave, Austria	Seeman, 1981
Malachite, $Cu_2(CO_3)(OH)_2$	Copper Queen Cave, Arizona	Ransome, 1904
Marcasite, FeS_2	Dachstein-Mammut Cave, Austria	Seeman, 1981
Mbobomkulite, $(Ni,Cu)Al_4(NO_3,SO_4)_2(OH)_{12} \cdot 3H_2O$	Mbobomkulu Cave, South Africa	Martini, 1980
Melanterite, $FeSO_4 \cdot 7H_2O$	Soldiers Cave, California	Rogers, 1976
Mellite, $Al_2[C_6(COO)_6] \cdot 16H_2O$	Romanelli Cave, Italy	Garavelli and Quagliarella, 1974
Metacinnabar, HgS	Gaudaskoy Cave, Russia	Lazarev and Philenko, 1976
Metatyuyamunite, $Ca(UO_2)_2(VO_4)_2 \cdot 3H_2O$	Spider Cave, New Mexico	Polyak and Mosch, 1995
Mimetite, $Pb_5(AsO_4)_3Cl$	Bisbee Cave, Arizona	Graeme, 1981
Minyulite, $KAl_2(PO_4)_2(OH,F) \cdot 4H_2O$	Boon Cave, South Africa	Martini and Kavalieris, 1978
Mirabilite, $Na_2SO_4 \cdot 10H_2O$	Crystal Cave, Kentucky	Bennington, 1959
Mitridatite, $Ca_2Fe_3(PO_4)_3O_2 \cdot 3H_2O$	Boons Cave, South Africa	Martini & Kavalieris, 1978
Monetite, $CaHPO_4$	Aggteleker Cave, Hungary	Sztrókay, 1959
Monohydrocalcite, $CaCO_3 \cdot H_2O$	Eiben Cave, Germany	Fischbeck and Müller, 1971
Montgomeryite, $Ca_4MgAl_4(PO_4)_6(OH)_4 \cdot 12H_2O$	Et-Tabun Cave, Isreal	Goldberg and Nathan, 1975
Montmorillonite, $(Na,Ca)_{0.3}(Al,Mg)_2Si_4O_{10}(OH)_2 \cdot nH_2O$	Westdriefontein Cave, South Africa	Martini and Kavalieris, 1978
Mundrabillaite, $(NH_4)_2Ca(HPO_4)_2 \cdot H_2O$	Petrogale Cave, Australia	Bridge and Clarke, 1983
Natrojarasite, $NaFe_3(SO_4)_2(OH)_6$	Jungle Pot Cave, South Africa	Martini 1983

Mineral	Cave	Reference
Natron, $Na_2CO_3 \cdot 10H_2O$	Saltzburg Shaft, Austria	Seeman, 1984
Nesquehonite, $Mg(HCO_3)(OH) \cdot 2H_2O$	Moulis Cave, France	Gèze and others, 1956
Newberyite, $MgHPO_4 \cdot 3H_2O$	Petrogale Cave, Australia	Bridge, 1977
Niahite, $(NH_4)(Mn,Mg,Ca)PO_4 \cdot H_2O$	Niah Cave, Malaysia	Bridge and Robinson, 1983
Nickelalumite, $(Ni,Cu)Al_4[(SO_4),(NO_3)_2](OH)_{12} \cdot 3H_2O$	Mbobomkulu Cave, South Africa	Martini, 1980
Niter, KNO_3	Goyder Cave, Australia	Mawson, 1930
Nitrammite, NH_4NO_3	Nickajack Cave, Tennessee	Shepard, 1857
Nitratine, $NaNO_3$	Goyder Cave, Australia	Mawson, 1930
Nitrocalcite, $Ca(NO_3)_2 \cdot 4H_2O$	Kartchner Caverns, Arizona	Hill and Buecher, 1992
Nitromagnesite, $Mg(NO_3)_2 \cdot 6H_2O$	Great Cave, Kentucky	Larsen, 1921
Olivenite, $Cu_2(AsO_4)(OH)$	Tintic Cave, Utah	Tower and Smith, 1899
Oxammite, $(NH_4)_2C_2O_4 \cdot H_2O$	Petrogale Cave, Australia	Bridge, 1977
Palygorskite, $(Mg,Al)_2Si_4O_{10}(OH) \cdot 4H_2O$	Broken Hill Cave, New Zealand	Lowry, 1964
Parahopeite, $Zn_3(PO_4)_2 \cdot 4H_2O$	Hudson Bay Cave, Canada	Walker, 1918
Phosphammmite, $(NH_4)_2HPO_4$	Petrogale Cave, Australia	Bridge, 1977
Phosphosiderite, $FePO_4 \cdot 2H_2O$	Westdriefontein Cave, South Africa	Martini, 1978
Pisanite, $(Fe,Cu)SO_4 \cdot 7H_2O$	Iron Canyon Cave, Nevada	Roberts and Arnold, 1965
Plattnerite, PbO_2	Bisbee Cave, Arizona	Graeme, 1981
Potassium alum, $KAl(SO_4)_3 \cdot 12H_2O$	Ruatapu Cave, New Zealand	Cody, 1978
Purpurite, $MnPO_4$	Gunong Keriang Cave, Malaysia	Jones, 1965
Pyrite, FeS_2	Dachstein-Mammut Cave, Austria	Seeman, 1981
Pyrolusite, MnO_2	Hall Cave, South Africa	Martini, 1993
Pyrrhotite, FeS	Playa Pájaro Cave, Puerto Rico	Kaye, 1959
Quartz, SiO_2	Wind Cave, South Dakota	White and Deike, 1962
Rhodochrosite, $MnCO_3$	Andalgala Cave, Argentina	Shaub, 1962
Romanechite, $BaMn_9O_{16}(OH)_4$	Jasper Cave, South Dakota	Moore, 1981
Rosasite, $(Cu,Zn)_2(CO_3)(OH)_2$	Bisbee Cave, Arizona	Graeme, 1981
Sabieite, $(NH_4)Fe(SO_4)_2$	Lone Creek Fall Cave, South Africa	Martini, 1983
Sampleite, $NaCaCu_5(PO_4)_4Cl \cdot 5H_2O$	Jingemia Cave, Australia	Bridge, 1978
Sasaite, $(Al,Fe)_{14}(PO_4)_{11}(SO_4)(OH)_7 \cdot 83H_2O$	Westdriefontein Cave, South Africa	Martini, 1978
Schertelite, $(NH_4)_2MgH_2(PO_4)_2 \cdot 4H_2O$	Chaos Cave, South Africa	Martini and Kavalieris, 1978
Scholzite, $CaZn_2(PO_4)_2 \cdot 2H_2O$	Island Ford Cave, Virginia	Dietrich, 1960
Sepiolite, $Mg_4Si_6O_{15}(OH)_2 \cdot 6H_2O$	Zbrazov Cave, Slovakia	Padera and Povondra, 1964
Shattuckite, $Cu_5(SiO_3)_4(OH)_2$	Bisbee Cave, Arizona	Graeme, 1981
Smithsonite, $ZnCO_3$	Herman Smith Cave, Illinois	Bradbury, 1959
Spangolite, $Cu_6Al(SO_4)(OH)_{12}Cl \cdot 3H_2O$	Blanchard Cave, New Mexico	Hill, 1976

Spencerite, $Zn_4(PO_4)_2(OH)_2 \cdot 3H_2O$	Hudson Bay Cave, Canada	Walker, 1918
Sphalerite, ZnS	Herman Smith Cave, Illinois	Bradbury, 1959
Stercorite, $H(NH_4)Na(PO_4) \cdot 4H_2O$	Petrogale Cave, Australia	Bridge, 1977
Stibnite, Sb_2S_3	Tyuya-munun Cave, Russia	Tsykin and Tsykina, 1979
Stilpnomelane, $KMg_2Fe_6(Si,Al)_{12}O_{15}(OH)_{12}$	Ferrata Cave, Italy	Hill and Forti, 1986
Strengite, $FePO_4 \cdot 2H_2O$	Westdriefontein Cave, South Africa	Martini, 1978
Struvite, $(NH_4)MgPO_4 \cdot 6H_2O$	Boon Cave, South Africa	Martini and Kavalieris, 1978
Sulfur, S	Cottonwood Cave, New Mexico	Davis, 1973
Swaknoite, $Ca(NH_4)_2(HPO_4)_2 \cdot H_2O$	Arnhem Cave, Namibia	Martini, 1991
Sylvite, KCl	Abjaterskep Cave, South Africa	Martini and Kavalieris, 1978
Syngenite, $K_2Ca(SO_4)_2 \cdot H_2O$	Murra-el-elevyn Cave, Australia	Bridge, 1977
Talmessite, $Ca_2Mg(AsO_4)_2 \cdot 2H_2O$	Luceram Cave, France	Feraud and others, 1976
Taranakite, $(K,NH_4)Al_3(PO_4)_3(OH) \cdot 9H_2O$	Pig Hole Cave, Virginia	Murray and Dietrich, 1956
Tarbuttite, $Zn_2(PO_4)(OH)$	Broken Hill Cave, Zambia	Spencer, 1908
Taylorite, $(K,NH_4)_2SO_4$	Murra-el-elevyn Cave, Australia	Bridge, 1975
Tenorite, CuO	Calumet Cave, Arizona	Ransome, 1904
Thermonatrite, $Na_2CO_3 \cdot H_2O$	Salzburg Shaft, Austria	Seeman, 1984
Thenardite, Na_2SO_4	Wind Cave, South Dakota	White and Deike, 1962
Tinticite, $Fe_4(PO_4)_3(OH)_3 \cdot 5H_2O$	Tintic Cave, Utah	Stringham, 1946
Todorokite, $(Mn,Ca)Mn_3O_7 \cdot H_2O$	Jingemia Cave, Australia	Bridge, 1978
Tridymite, SiO_2	Cango Cave, South Africa	Martini, 1993
Tschermigite, $(NH_4)Al(SO_4)_2 \cdot 12H_2O$	Lone Creek Fall Cave, South Africa	Martini, 1983
Turanite, $Cu_5(VO_4)_2(OH)_4$	Tyuya-munun Cave, Russia	Chirvinsky, 1925
Tyuyamunite, $Ca(UO_2)_2(VO_4)_2 \cdot 6H_2O$	Tyuya-munun Cave, Russia	Chirvinsky, 1925
Urea, $CO(NH_2)_2$	Wilgie Mia Cave, Australia	Bridge, 1973
Uricite, $C_5H_4N_4O_3$	Dingo Donga Cave, Australia	Bridge, 1974
Vanadinite, $Pb_5(VO_4)_3Cl$	Havasu Canyon Cave, Arizona	McKee, 1930
Variscite, $AlPO_4 \cdot 2H_2O$	Drachen Cave, Hungary	Machatschki, 1929
Vivianite, $Fe_3(PO_4)_2 \cdot 8H_2O$	Castellana Cave, Italy	Balenzano and others, 1974
Weddellite, $Ca(C_2O_4) \cdot 2H_2O$	Petrogale Cave, Australia	Bridge, 1977
Whewellite, $CaC_2O_4 \cdot H_2O$	Parakiet Cave, Namibia	Martini, 1993
Whitlockite, $Ca_9(Mg,Fe)H(PO_4)_7$	El Chapote Cave, Mexico	Pérez and Wiggen, 1953
Witherite, $BaCO_3$	Lilburn Cave, California	Rogers and Williams, 1982
Woodhouseite, $CaAl_3(PO_4)(SO_4)(OH)_6$	Jade Lotus Cave, China	Wang, 1982

The abrupt transition within a cave entrance zone leads to great biologic variety because the population consists of all species that are tolerant of the twilight zone. Shown here is La Cueva de Camuy, Puerto Rico.

5: Behavior and Products of Cave Microorganisms

THE PLANT LIFE of caves is made up of species that can live in total darkness. If any plant in a cave contains chlorophyll, the chlorophyll obviously cannot be energized by light rays from the sun. With a few exceptions noted below, none of the plants in the perpetually dark zone contains chlorophyll. The chief plant life there consists of bacteria (including actinomycetes) and fungi.

With regard to their mode of life within the cave, these microorganisms are divided into *hetereotrophs* and *autotrophs*. Each group requires carbon compounds for its nutrition, but the heterotrophs are unable to synthesize them and must obtain organic material by consuming the remains of other plants found in the cave. The autotrophs, on the other hand, are able to build the organic substances essential for life directly from inorganic raw materials. Most autotrophs that live on the surface contain chlorophyll and derive the energy for food synthesis from the sun. In the cave environment, however, certain autotrophic bacteria, known as *chemoautotrophs*, are able to obtain all the energy they need from the transformation of certain minerals to different ones.

These chemoautotrophs play an important role in the food chains of some caves. After synthesizing food from inorganic material, they serve as food for such small animals as protozoans, which in turn are eaten by larger animals. Thus, the chemotrophs theoretically can serve as the basic food supply for the life of the entire cave.

Heterotrophic bacteria and fungi break down waste material deposited by cave-dwelling animals as well as organic material brought in by flowing water or visiting animals. In so doing they perform two functions. First, they act as scavengers; second, they release chemical compounds for further use as nutrient material for other organisms. Heterotrophic bacteria cannot exist in areas devoid of organic material, but autotrophic bacteria can. Life can therefore go on in a cave that is sealed from the surface, provided its flora includes a few autotrophic microorganisms.

Characteristics of the Microflora

SINCE A CAVE is normally connected to the surface by an entranace, microorganisms found in the dark zone are similar to species found in the surface soil. The spores by means of which they propagate are so small that they can readily be carried deep into the cave by percolating soil water, as well as by currents of air, by streams of water flowing into the cave, or by animals. When they have settled in a suitable environment, the spores germinate and develop into mature forms of the species.

Demonstrations in English caves have shown that the mere passage of explorers into and out of a cave can cause extreme contamination by certain types of bacteria that had not previously existed in the cave. Sterile petri dishes placed in a virgin cave area and then cultured have proved to be free from bacteria of certain outside species, but when cultured after a party of ten or twelve people had passed through this same area, they contained many of these bacteria.

To determine the extent to which a cave has been invaded by heterotrophic forms brought in from the surface, either by air or by water, we need only take samples at a great many places in a cave and culture them for molds. The French microbiologist Victor Caumartin has pointed out that molds, and the microorganisms that accompany them, penetrate no farther into a cave than tiny fragments of surface plants that have been carried in by natural agencies. In passages beyond those containing the innermost molds, the microscopic plant life is chiefly autotrophic. The presence of molds may therefore define the contaminated areas. At the boundaries between contaminated and uncontaminated areas, the heterotrophic surface bacteria that break down organic compounds, and the autotrophic underground bacteria that synthesize these compounds overlap and compete with one another.

Among the autotrophic microorganisms commonly found in caves are the chemoautotrophic iron bacteria. This is not surprising, because most caves contain everything these bacteria need in order to live, including an abundance of moisture and of iron compounds, and a sufficient quantity of certain indispensable trace elements. Caumartin has shown how these resources are utilized by the iron bacterium *Perabacterium spelei*. This species is anaerobic—that is, it requires no free oxygen—and it can fix nitrogen obtained from the air. It derives the carbon it needs from iron carbonate in the walls of the cave. Decomposition of the iron carbonate supplies the energy required for the bacterium's metabolism. This process liberates ferrous ions that are oxidized to produce the ferric mineral goethite [FeO(OH)], which is the brown pigment of cave silt.

Perabacterium spelei is considered by Caumartin to be the first link in a food chain that requires nothing from outside the cave. Experiments with the cave-dwelling amphipod *Niphargus* have demonstrated that it can subsist in tanks with no food other than clay containing iron minerals and iron bacteria. If no clay is available to young specimens of *Niphargus*, even if organic food is supplied, the young die before achieving their second molt. The *Niphargus* also soon die if the supply of iron minerals in the clay becomes exhausted. *Perabacterium spelei* can therefore serve as food for the *Niphargus* in a cave that is entirely devoid of the chlorophyll-bearing plant materials usually considered essential to animal life on the surface.

Energy is released too by the slow process of transformation from one clay mineral to another in the sediment that partly fills caves. Muscovite, an insoluble mineral derived from beds of shale associated with the limestone, is the principal original constituent of the clay-sized fraction of cave sediment. This muscovite formed in the shale when it was under great pressure in the Earth. As a consequence, muscovite is unstable in the low-pressure cave environment, and it slowly changes there into calcium-montmorillonite, another clay mineral. We suggest that this and similar transformations of clay minerals may be important energy sources for cave chemoautotrophs.

The peculiar odor, suggesting damp earth or moldiness, so characteristic of some caves, is produced by cave actinomycetes, which are moldlike filamentous bacteria. Some actinomycetes synthesize carotene, a common pigment of certain cave-dwelling insects. The brownish color often found in droplets of water coating cave walls is also caused by this pigment.

Algae occur in caves, and some of them, contrary to a widely held belief, can live heterotrophically in darkness in the presence of organic material. It has usually been assumed that all algae contain chlorophyll, and therefore can synthesize nutrients only in the presence of sunlight. Recently, however, George Claus has found several species of algae that grow in total darkness in Béke Cave and Baradla Cave in northern Hungary. Chlorophyll is present in some of the species, but absent in others.

Sulfur bacteria are found mainly in caves that have developed near shaly rock strata containing such sulfide minerals as pyrite. Because most caves contain some sulfur compounds, almost every cave probably contains sulfur bacteria. Investigators have found that bacteria utilizing sulfur compounds produce vitamins essential for animal growth. These are the familiar B vitamins, all of which have been isolated from ordinary sulfur bacteria. Most organisms synthesize the B vitamins from amino acids found in chlorophyll-bearing plants that require sunlight. Sulfur bacteria can synthesize them in caves in total darkness.

Microorganisms are themselves eaten by cave insects as a source of amino acids and other materials required for growth. Some microorganisms live in the intestines of cave insects. Here they aid digestion by secreting enzymes that help to convert food to a form that can be absorbed. Cave microorganisms therefore play an important role in enabling cave animals to survive and grow on what would otherwise be an inadequate diet, and to do this in an environment entirely without sunlight.

Saltpeter

A MATERIAL that owes its origin in part to microorganisms is cave saltpeter, which was extensively mined to make gunpowder during the War of 1812 and the Civil War. This highly soluble material, consisting mainly of nitrocalcite [$Ca(NO_3)_2 \cdot 4H_2O$], impregnates silt in many caves. Miners processed the saltpeter-bearing silt, or "petre dirt," by leaching it with water and then boiling the liquor with wood ashes to convert it to niter [KNO_3], with which to make gunpowder.

The bacterium *Nitrobacter agilis*, responsible for the final stage of forming the nitrate of saltpeter from nitrogenous organic material, is common in saltpeter caves. A puzzling aspect of the problem, however, is that some of the caves richest in saltpeter contain no obvious accumulations of such organic material as bat guano. W.H. Hess proposed an explanation for cave saltpeter in 1900, and Carol Hill has developed it more fully in recent years. They suggest that the nitrate ions originate at the surface by the decay of plant material in the soil. Nitrate then migrates downward in low concentration in percolating water. In this hypothesis, when the water encounters a dry cave passage, it evaporates, and the nitrate ions become more concentrated. In effect, the circulating cave air draws the water out until the concentration of saltpeter constituents reaches the levels now found in the caves.

In the first edition of this book, we originated an alternative hypothesis based on the fact that during the War of 1812 the principal source of saltpeter other than caves was the soil around houses and under their floors. Before the days of modern plumbing, it was in this soil that nitrogen compounds from human waste accumulated. We suggest that the ultimate source of the raw material for saltpeter in caves may be the urine of the cave rat *Neotoma magister*. These animals leave a trail of urine droplets and musk in the dark zone of caves, which helps them to find their way from the surface back to their nests. In some dry western caves, where cave rats have been crossing an area of limestone for thousands of years, thick accumulations of dehydrated urine called *amberat* have formed. On silt, however, or in wet caves, bacterial action may break down the urea and convert it to cave saltpeter.

Manganese and Iron Minerals

MICROORGANISMS also play a part in the origin of the black manganese deposits commonly present in caves. The manganese minerals form sootlike layers that cover the walls of certain passages or coat cobbles in cave streams. In most limestone caves, as in Weber Cave, Iowa, these layers are composed of the mineral birnessite [$CaMn_7O_{14} \cdot 3H_2O$]. In Jasper Cave, South Dakota, however, where the ground water contains a higher than average amount of barium, the mineral is romanechite [$BaMn_9O_{16}(OH)_4$]. The deposits tend to be layered like stalactites. They consist of crystals so extremely minute that they are almost beyond the limit of resolution of the x-ray diffraction technique normally used for identifying such minerals.

Limestone and cave water ordinarily contain enough manganese to provide a source for the manganese in the cave deposits. The common occurrence of the black material on stream cobbles a short distance into caves from the surface suggests that an increase in alkalinity caused by a greater bicarbonate concentration creates a favorable chemical environment for manganese deposition. The dissolved manganese is held in solution in the cave water in a chemically reduced form, either as Mn^{2+} ions or as part of a soluble complex organic molecule. Specialized bacteria such as *Clonothrix fusca* and *Leptothrix discophora*, which are known to precipitate manganese in domestic water-supply systems, may live in the caves and cause the deposits to form there. The bacteria derive energy from the oxidation of the manganese from Mn^{2+} to Mn^{4+} and free it from the complex molecules, thus causing the water to become supersaturated with manganese near the bacterial colonies. This causes the birnessite or romanechite to be deposited.

Scanning electron micrograph of a black manganese coating on a stream cobble from well aerated, slightly alkaline water in Matts Cave, West Virginia. This porous material, enlarged 10,000 times, is composed of the mineral birnessite. The birnessite grains may have been deposited around the sheaths of bacteria such as *Leptothrix discophora*, which have about the diameter of the pores in the coating.

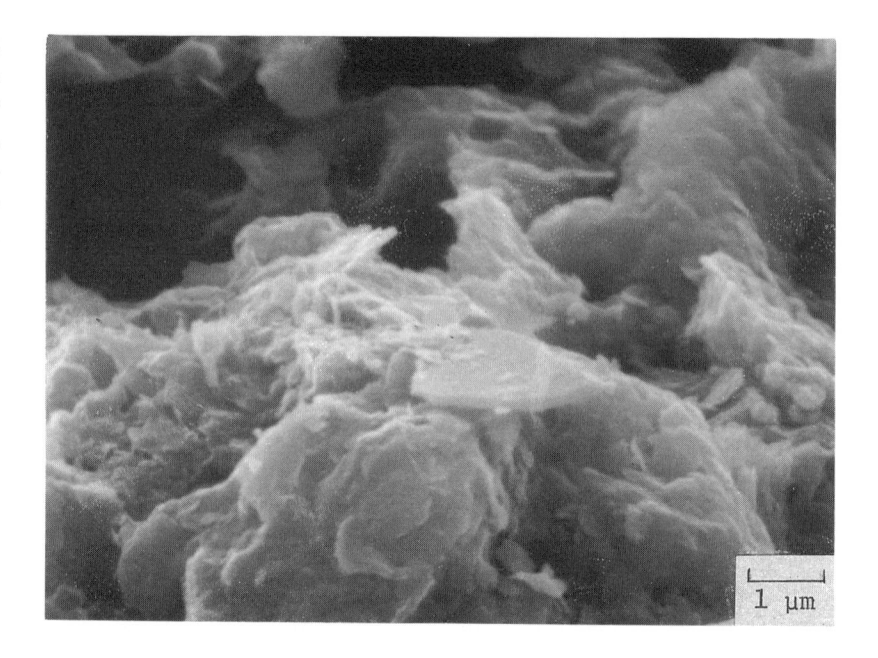

Black manganese minerals intermix with brown iron minerals in some caves, and in others, the iron minerals occur alone. Bacteria have been intermediaries in all of these deposits.

In Sandia Cave, New Mexico, for example, a layer of yellow ocher containing the bacterium *Leptothrix ochracea* covers the entrance room of the cave. It consists mostly of the iron-mineral goethite [FeO(OH)]. The limestone of Sandia Cave directly overlies granite that provides a source for the iron. When the land surface was higher, water that percolated through the granite moved up into a cave pool. When the iron-bearing water met free oxygen at the surface of the pool, the iron bacteria deposited the goethite.

Frederick Luiszer has studied a nearly identical case at Cave of the Winds, Colorado, except that both iron and manganese minerals were deposited there. Sediment in the cave contains a layer of iron minerals 13 centimeters thick, overlain by a layer of manganese minerals 5 centimeters thick, overlain in turn by several meters of clay and silt.

The nearby Ouray Spring, which issues from granite and deposits iron minerals, and the closer Shoshone Spring, which issues from limestone and deposits manganese minerals, are used as evidence to explain the deposits in Cave of the Winds.

Luiszer infers that a slow downcutting of surface valleys gradually changed the relationship between the ground water, the granite, and the limestone conduits. This downcutting caused a sequential change in conditions at Cave of the Winds that led to deposition of the iron, then the manganese, then the clay and silt, and finally drained the cave. The cave minerals contain fossil bacteria, and the surface springs contain living mucilaginous iron and manganese bacteria.

Moonmilk

LABORATORY CULTURES made from the water found on some cave deposits reveal the presence of bacteria and other microorganisms that may play a part in the construction of certain calcareous mineral deposits and in the disintegration of the limestone wall rock.

Microorganisms have been shown, for example, to play a major role in the origin of a curious cave material known as *moonmilk*. This is a soft, white, claylike substance present on the walls of many caves. Its name comes from the German-speaking part of Switzerland, where in the 16th Century people called it Monmilch, Mon being an old spelling for Mond, the present German word for moon. The people then believed that this underground white material was formed by the rays of the Moon. In the 16th Century, it was generally thought that the nightime dew was produced by the Moon. At that time, light from celestial bodies was believed to gain substance as it passed through rock, thereby producing metallic ores, for example, gold from the Sun and silver (and moonmilk) from the Moon.

The mineralogy of moonmilk has been intensively studied in Europe and North America during the past 30 years. We now know that the microscopic grains in the moonmilk of limestone caves consist mainly of calcite [$CaCO_3$]. But in relatively warm caves where the wall rock contains appreciable quantities of magnesium as a constiuent of the mineral dolomite, the grains may consist of any of the following magnesium minerals: hydromagnesite [$Mg_5(CO_3)_4(OH)_2 \cdot 4H_2O$]; magnesite [$MgCO_3$]; huntite [$CaMg_3(CO_3)_4$]; nesquehonite [$Mg(HCO_3)(OH) \cdot 2H_2O$]; and dolomite [$CaMg(CO_3)_2$].

When the mineral constituents of moonmilk are removed by dissolving it in a weak acid, an abundant organic residue remains, which consists chiefly of such bacteria as *Macromonas bipunctata*, along with actinomycetes and algae. This microflora may assist in breaking down the minerals of the wall rock, and it aids in their conversion to the solids contained in the moonmilk.

Scanning electron micrograph of a calcite rod in moonmilk from Liburn Cave, California, showing a diagonal surface texture that follows a diagonal internal calcite crystal structure, typical in moonmilk but uncommon for calcite.

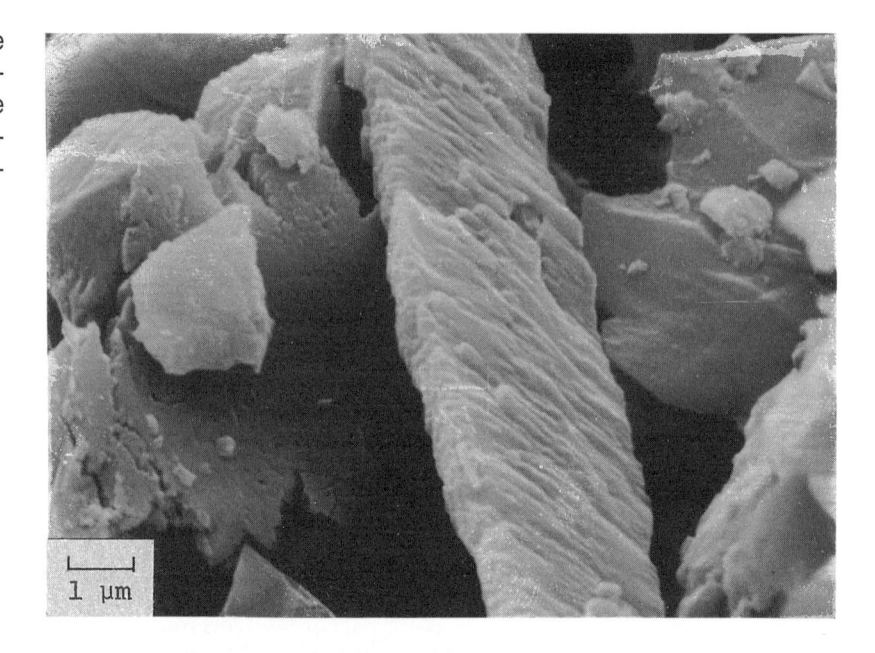

A calcite rod in moonmillk from Crystal Cave, California, surrounded by finer rods. The calcite is believed to be nucleated by actinomycetes—moldlike bacteria.

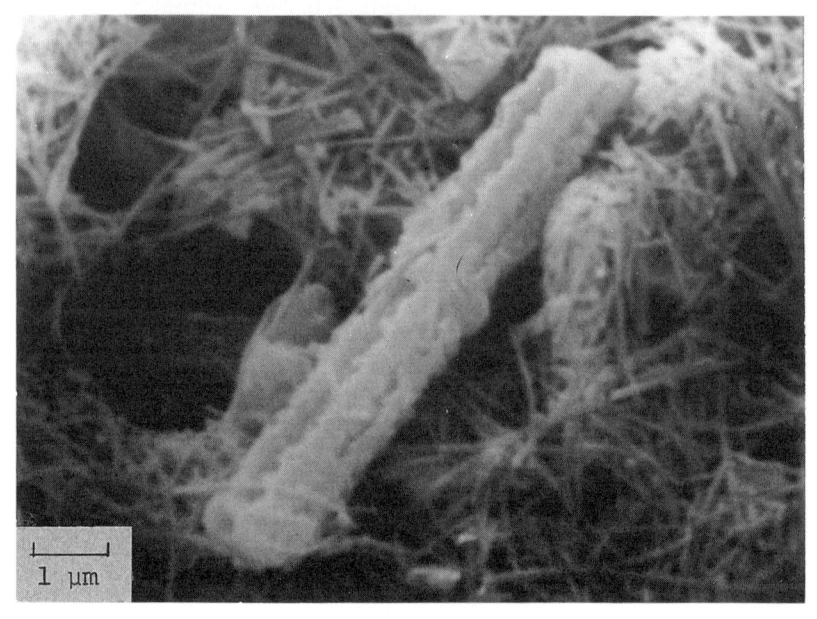

Moonmilk from Lilburn Cave, California, composed of the second most common moonmilk mineral, hydromagnesite, showing typical monoclinic crystal plates.

Moonmilk from Carlsbad Cavern, New Mexico, composed of hydromagnesite (large tablets) and huntite (smaller felted grains).

The larger mineral bodies in calcite moonmilk have a distinctive surface sculpture that can be seen under the scanning electron microscope. The bodies consist of rods of calcite with an average size of 1 x 8 micrometers. A diagonal grain is impressed on the surfaces of the rods, and parallel ridges commonly trend along the lengths of the rods superimposed on the diagonal grain. The combined effect produces bodies somewhat resembling ears of corn.

The diagonal grain is aligned with the crystal structure, as can be seen through an optical microscope with polarized light. Because the crystal structure of calcite normally parallels the long dimension of calcite crystals, the grains in calcite moonmilk were once erroneously identified and named as a separate new mineral, "lublinite."

In many samples, the rods are enmeshed in a net of calcite filaments about 0.1 micrometer in width. These filaments are believed to have been associated with microbes that serve as nuclei for growth of the calcite bodies. An inclined crystal face of a seed crystal is nucleated by the microorganism and subsequent growth leads to the unusual crystal orientation and surface sculpture of the moonmilk grains.

Most moonmilk occurs where water may reasonably be inferred to move through the substance to its surface where deposition takes place by loss of carbon dioxide. The life processes of individual microorganisms cause a microvariation of the chemical environment that leads to deposition of discrete mineral grains, rather than to a more solid speleothem such as cave coral.

In a specimen of fresh calcite moonmilk from Caverns of Sonora, Texas, imaged by transmission electron microscopy, several of the thinnest (and shortest) mineral filaments have round organic bodies 0.1 micrometers in diameter at one end. We suggest that this offers an explanation for the small size of moonmilk grains. The organism nucleates the rod from one end only, until it divides. Subsequent thickening and sculpturing of the rod then takes place mainly by inorganic mineral overgrowth.

More research is needed to determine the energy source of the microorganisms in moonmilk. The snow-white rather than brown color of most moonmilk suggests that oxidation of iron is not the source.

In 1994, Rudolph Reinbacher studied freshly forming moonmilk in Mondmilchloch, Switzerland, the type locality for the substance. The moonmilk there is calcite with a color (wet) in the Munsel color system of very pale orange (10YR8/2). A sample stored in a jar in dim light for several months acquired brown, gray, and green spots, presumably from microbial contamination. This growth proves that nutrients re-

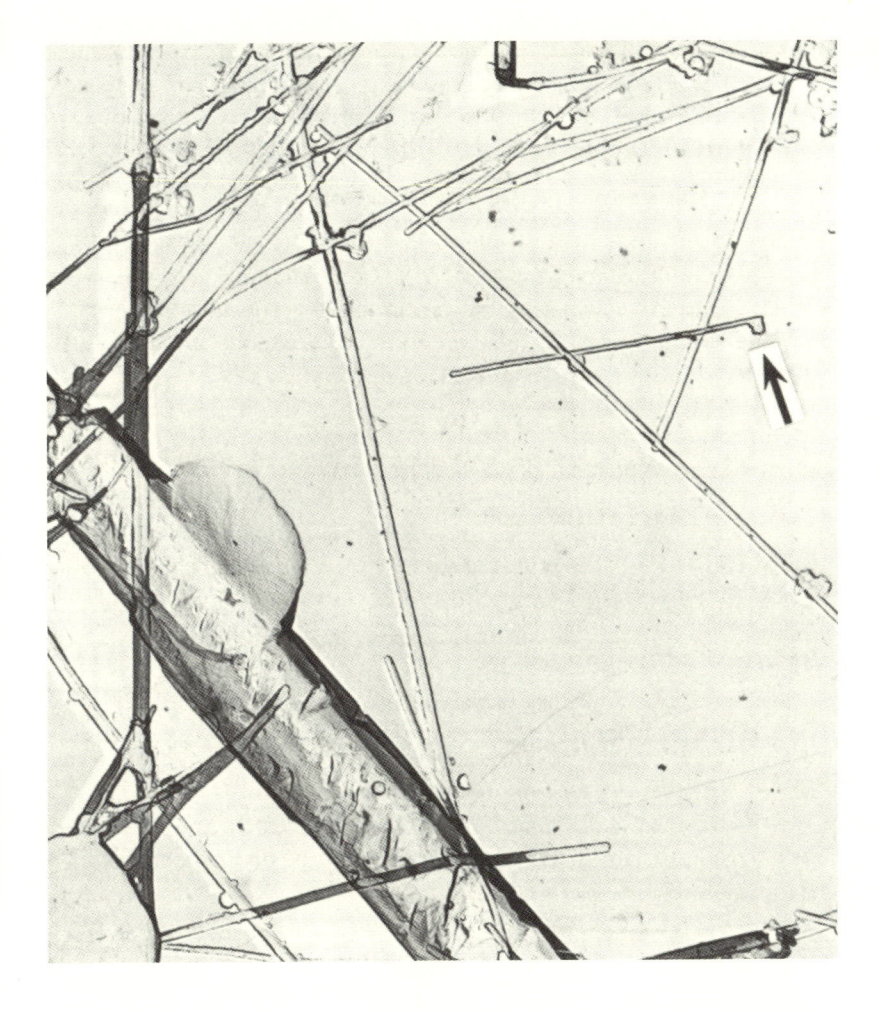

Transmission electron micrograph of calcite rods and organic bodies in moonmilk from Caverns of Sonora, Texas. The small cell at the end of a slender rod at the arrow at center right suggests a mechanism for restricting the size of moonmilk grains. The cell causes rod growth from one end, and it forces the calcite to precipitate in a nontypical diagonal alignment across the rod. The rod lengthens until the cell divides, at which time it stops. After that, the rod gets larger in diameter by inorganic precipitation, but it does not get significantly longer. The image is 8 micrometers wide.

mained in the sample and suggests that soluble organic compounds from the soil provide the energy for the microorganisms that control the growth of this strange substance.

Medical Use of Cave Actinomycetes

CAVE ACTINOMYCETES have been the subject of much research as a possible source of antibiotics. Several expeditions into caves in Central America, South America, and the United States have been conducted to collect cave silt that contains actinomycetes and molds. From these, scientists hope to obtain new and powerful antibiotics.

In the 16th and 17th centuries, long before modern "miracle drugs" were dreamed of, physicians used dried moonmilk from European caves as a dressing for wounds. They did so primarily because this substance would stop bleeding and act as a dehydrating agent, but they also believed it had curative qualities. Now that we know that moonmilk contains actinomycetes, and some actinomycetes possess antibiotic properties, we see that the early use of this cave material in medicine may have had a valid scientific basis, even though the early physicians did not know what it was.

It is of passing interest to note here that speleologists testify that sometimes when they enter a cave while suffering from a cold, they find that after they have been underground for several hours their symptoms largely disappear. A probable explanation for most of these cases is that deep inside caves the air is almost free from pollen, and the clean air would alleviate symptoms caused by some allergies. It is possible, however, that in some cases the cold victim obtains relief by inhaling an unknown cave product that may someday be used medically in treating common colds.

Harmful Microorganisms

NOT ALL SPECIES of cave microflora are beneficial to people. As noted above, certain bacteria cause the breakdown of organic material—a welcome sanitary process when the material broken down is organic waste. But the same bacteria also disintegrate what might have been especially interesting remains. Countless vertebrates, including humans, have been buried in caves, yet it is rare to find anything more than their bones, because the cave bacteria have usually caused complete decay of all other parts. The only exceptions to this are in certain caves with extremely low humidity; in these the microflora is nearly inactive and in especially dry areas it is sometimes possible to find desiccated bodies. Thus, in recent years, a sequence of extinct marsupial species with skin and fur intact has been studied from the caves under the Nullarbor Desert of Australia. Also, in drier parts of caves in Mammoth Cave National Park, Kentucky, desiccated human bodies over 2,000 years old have been found.

Several pathogenic microorganisms are also known to exist in caves. One of these *Histoplasma capsulatum* is responsible for a disease known as histoplasmosis, whose symptoms and effect are similar to those of tuberculosis. In recent years, cave explorers in the United States, Central America, Australia, and South Africa have contracted histoplasmosis by inhaling the spores of *Histoplasma capsulatum*. Outbreaks of histoplasmosis from noncave sources have been reported in the Ohio and Mississippi Valleys. The spores have been detected in chicken coops and roosting areas of other birds where dried droppings accumulate, as well as in caves. Symptoms of histoplasmosis include loss of weight, anemia, fever, coughs, and severe chest pain. The therapeutic agent administered for severe cases is Amphotexicin-B.

Relation of Microorganisms to Cave Food Chains

MICROORGANISMS are an extremely important constituent of the cave environment. They are involved in the development of cave deposits such as moonmilk, in the production of food for cave animals, and in the breakdown of organic material in the cave.

Ideally, a nearly closed ecologic system could exist in the dark zone of a cave where the energy required for metabolism is derived from minerals in the wall rock and in sediment on the floor. The basic food cycle in such a cave setting would depend on chemoautotrophic bacteria. These bacteria could serve as nutrient material for cave-dwelling animals, with no organic input from outside the cave.

Most caves that contain established communities of cave organisms, however, are ecologically similar to abyssal depths in the ocean, where the sunlight required for supporting the green plants on which the lives of the surface fauna depend is completely absent. In each case, bacterial decomposers rely on a flow of organic material from sunlit areas. In caves, this can be from dissolved organic matter in drip water, from plant remains in cave streams, or from organic substances in cave silt. Organic material is also obtained from the fecal matter of animals that periodically spend time outside the cave, such as bats, cave rats, and crickets.

Thus, the primary energy sources exploited by cave organisms are either minerals broken down by autotrophic bacteria, or surface-derived organic material utilized by heterotrophic bacteria. In the quantitatively more important heterotrophic case, the decomposer bacteria are eaten by protozoans that are eaten by such aquatic cave-dwelling animals as flatworms, isopods, and amphipods, that are eaten in turn by larger animals such as crayfish, salamanders, and fish. Finally, these aquatic forms release waste material that supports the heterotrophic bacteria that helped to initiate the chain, and the cycle is complete.

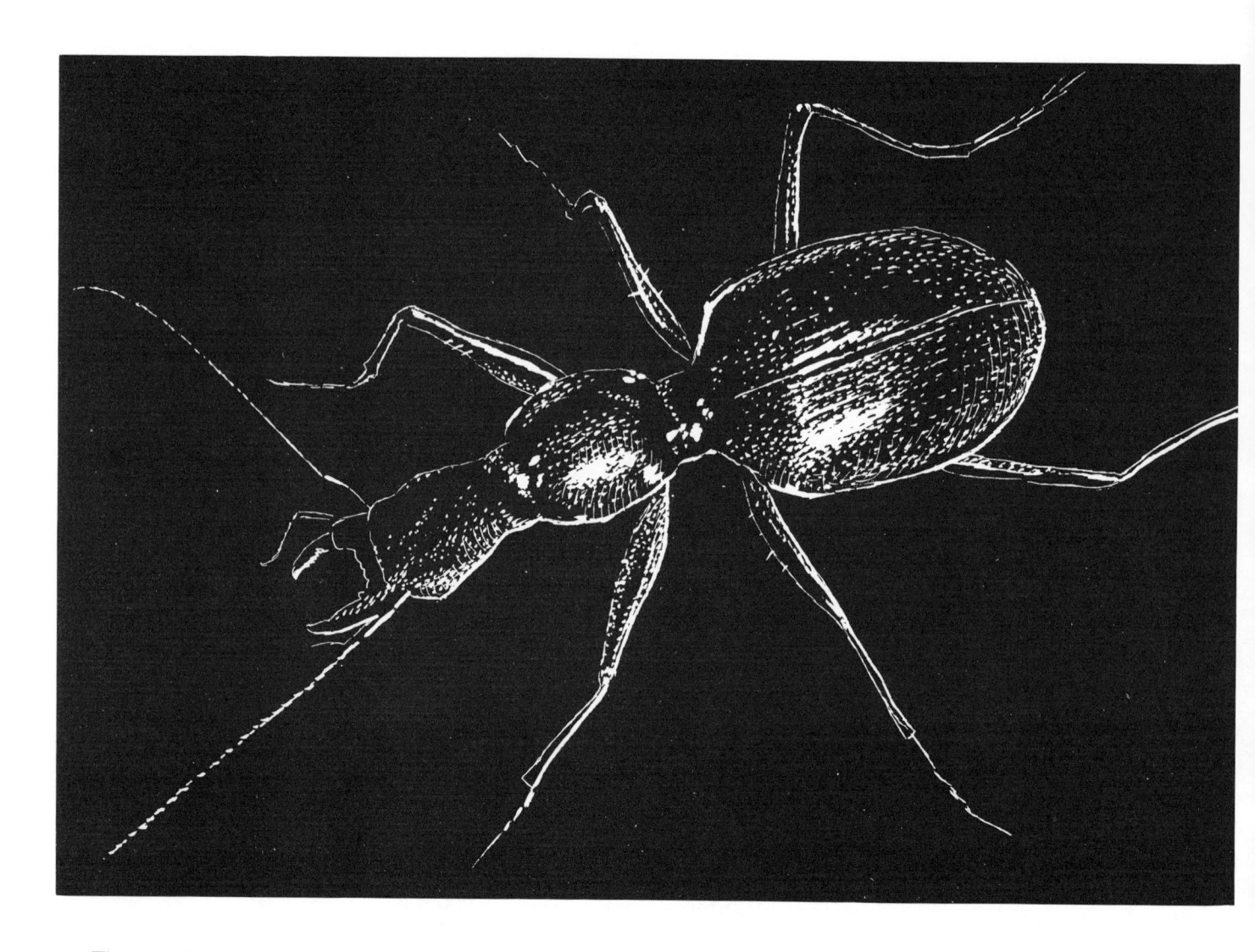

The cave beetle *Neaphaenops tellkampfi*, whose body is 5 millimeters long, not only lacks eyes but also does not even have an optic nerve.

6: Habits of Cave Animals

A STUDY OF CAVE LIFE involves not only the plants and animals themselves, but also their environment and the ways in which it affects them. The most obvious characteristic of caves is the prevailing utter darkness beyond the twilight zone that extends a short distance in from the entrance. A similar degree of darkness indeed surrounds parasites that live inside their host and the billions of animals that live deep below the surface of the sea. Such darkness, however, is entirely beyond the experience of organisms that live on the surface of the Earth. Outside a cave, even on the darkest night, people can see dim outlines after their eyes become accustomed to the dark, but well inside a cave they literally cannot see their hands before their faces. An experiment with photographic film gives striking proof of this complete absence of light: A film developed after being exposed for a week in the depths of a cave was completely blank.

Another characteristic feature of the cave environment, noted in Chapter 3, is its almost completely uniform climate. The atmosphere in a deep cave rarely changes, except for slight variations in barometric pressure and carbon dioxide content. The temperature remains virtually constant throughout the year, a factor especially beneficial to animals that cannot regulate their body temperatures.

An additional factor important to cave animal life is the nearly constant high relative humidity. In most caves water continually drips from the ceiling and the walls, and the drip water either accumulates in pools or flows away in streams. Cave walls are sometimes so moist that aquatic organisms can readily crawl over them. Small crustaceans and flatworms, that normally live only in pools or streams outside of caves, are found even on the ceilings of humid caves.

Other caves, however, are so dry and dusty that a person finds it hard to breathe in them when the dust has been disturbed. These dry caves are exceptional. Life is rare in such caves, and they are of interest chiefly to archeologists and paleontologists, who look in them for evidence of earlier habitation by people or extinct species of animals. Biologists are interested chiefly in caves with a moist atmosphere, for these are the only ones that now contain large numbers of living organisms.

Organisms that pass their entire lives in the inner part of a typical cave, then, must be adapted to continuous utter darkness, combined with high humidity and moderate temperature, both virtually constant throughout the year. The animals most thoroughly adapted to such conditions may also possess other special characteristics, such as extra-long feelers or antennae for finding their way about.

There are degrees of adaptation to cave life. Adaptation is relatively slight in animals that live in caves only during the winter or for even shorter periods. One convenient way to classify cave animals is according to how much of their time they spend in caves.

Degree of Adaptation to Caves

CAVES, ESPECIALLY THEIR ENTRANCE ZONES, are often inhabited temporarily by animals that usually live above ground but occasionally move into a cave for protection. Bears use caves for their long winter sleeps. Bats may remain in caves continuously throughout the winter when they are hibernating, but in the summer they rest in them only during the daylight hours. Skunks, raccoons, moths, mosquitoes, and even people use caves as wintering places or take refuge in them to avoid extreme heat, cold, or storms. Animals that do this are called *trogloxenes*, from the Greek words *troglos* (cave) and *xenos* (guest). Trogloxenes never complete their whole life cycle in a cave.

Other species of animals regularly live in the dark zones of caves, although they can and do survive outside caves, provided the environment is moist and dark. Earthworms are a good example, and some of the salamanders, beetles, and crustaceans found in caves are included in this category. They are known as *troglophiles*, from *troglos* (cave) and *phileo* (love). Some individual troglophiles may complete their entire life cycle in a cave, but other individuals of the same species live outside.

Last, there are forms that live permanently in the dark zone and are found exclusively in caves. They commonly have very reduced pigment and have either very small eyes or none at all. These forms are known as *troglobites*, from *troglos* (cave) and *bios* (life). They include the blind cave fishes *Amblyopsis spelaea* and *Tryhlichthys subterraneus*, such isopods as *Asellus propinqus*, and such salamanders as *Haideotriton wallacei* and *Typhlomolge rathbuni*. Troglobitic insects frequently have longer appendages and thinner shells than do related surface forms. All existing troglobites usually have evolved from troglophiles.

Another group of organisms that can live in caves but are not technically regarded as cave forms, are parasites, such as the ticks and lice that infest cave-dwelling animals. A few parasitic species, however, are classed as troglobitic because they are never found outside a cave. Bats in particular carry so many ticks and lice that, in caves where large numbers of bats assemble, these parasites rain continually from the ceiling. Parasites are also found in the gills of cave crustaceans and on the skins of cave fish and salamanders.

Some microscopic types of animal life, such as protozoans and rotifers, are found in cave streams. All cave streams examined for evidence of life have been found to contain some protozoans. A total of 33 different species were described from just two caves in Romania. Some forms were undoubtedly first carried in from the surface, and their presence in caves is accidental. All of these microscopic forms serve as food for larger aquatic cave animals.

Troglobites

WE ASSUME THAT WITH ADDITIONAL STUDY the bacteria now known only from caves will someday be found in other environments such as the surface soil. Therefore, such bacteria are excluded from the troglobite group. The simplest of the troglobites are some of the one-celled protozoans. Next in ascending order of complexity are the cave-dwelling flatworms known as planarians, which are rarely more than 2 centimeters in length. Flatworms are aquatic and feed by extending a tube from the center of the body. The cave forms are white and eyeless. Most of them have an adhesive organ in the middle of the front end, which enables them to hold fast to rocks and pebbles. Flatworms can, in addition, float through the water; they are found in large numbers feeding on the bodies of bats and other organisms that fall into the water. They have the remarkable ability to regenerate a new head or tail, and will form two complete organisms if cut in half. Little is known of the life cycle of cave planarians.

Snails and slugs are attracted to the moist environment of a cave. Several dozen species of snails are troglobites. They slither along walls and floors, feeding on tiny fragments of organic matter. Slugs have been seen invading the burrows of beetles, and possibly they feed on dead cave animals, but no troglobitic slugs are now known.

Crustaceans are among the more prominent and most abundant inhabitants of caves, with amphipods, isopods, freshwater shrimp, and crayfish predominating. Ostracods, resembling tiny clams, and copepods, one type of water flea, may be present in large numbers, although being microscopic they are rarely noted. Amphipods, which resemble small shrimp, are laterally compressed, whereas isopods, of which sow bugs constitute one type, are horizontally flattened. The amphipods and isopods feed on bacteria and fine-grained organic material in the streams and pools of the cave. If moisture is abundant, these small crustaceans crawl from the water and move about on the walls and floors. They have only a thin shell, but as long as their bodies remain moist, they can survive out of water.

Cave crayfish are not all colorless, and many are closely related to species that do not live in caves. *Orconectes pellucidus*, a form abundant in many caves of the southeastern United States, is about 15 centimeters long, or as large as the crayfish in surface streams. The cave crayfish, however, utilize their food much more slowly and can live twice as long as surface forms in water with a given content of dissolved oxygen. The cave crayfish's slow life processes facilitate its survival during times of diminished food supply.

Known Troglobites of North America

Class	Families	Species and subspecies
Tubellaria (flatworms)	4	24
Gastropoda (snails)	4	13
Crustacea	30	285
Diplododa (millipedes)	23	88
Chilopoda (centipedes)	5	9
Insecta	19	310
Arachnida (spider relatives)	33	174
Osteichthyes (bony fishes)	8	14
Amphibia	1	9
TOTAL	127	926

Cave crayfish and other cave-dwelling crustaceans occasionally make their way into the watecourses that flow into springs or wells.

Most species of millipedes in caves are similar to those found on the surface but *Scoterpes copei,* shown here, is troglobitic, never being found on the surface. It lacks pigment and is blind.

More than one farmer has been astonished to find a colorless crayfish in a bucket of well water.

Millipedes form another large group of organisms that require a moist environment and are common in caves. They burrow into cave silt and feed on humus and fungi. The troglobitic forms have less body pigment, thinner shells, longer legs and antennae, fewer eye facets, and longer sensory bristles than surface forms. In some of the cave forms, the shell is so thin that it will wrinkle in dry air. Millipedes, provided there is sufficient moisture, can live anywhere in a cave. Because they feed on anything organic, from bat guano to paper discarded by cave explorers, millipedes are widely distributed. Centipedes also frequent caves, but only a few supposedly troglobitic species are found in North America.

Earthworms burrow in the mud of many caves. They ingest the mud, extract nutrient material from it, and then pass the remainder through their digestive tracts. Their castings contain sufficient nutrients to provide food for other organisms that live in the mud. No truly troglobitic species of earthworms are known; those found in caves are also found in surface soils.

Visitors to caves are sometimes startled to observe small insects, rarely more than 5 millimeters long, springing for distances of more than a meter. These are the springtails or collembolans. Springtails live in most caves, although they frequently escape notice because of their small size. They feed on decaying organic material.

Many species of winged insects live in caves. These include various species of flies, gnats, and crane flies. Mosquitoes use caves for overwintering, and speleologists are occasionally surprised to find themselves bitten by mosquitoes because they have disturbed a swarm inside a cave. None of the winged insects are troglobites, but many are trapped in the webs of cave spiders and thus become important items in the cave food supply.

The glowworm *Arachnocampa luminosa*, which inhabits Waitomo Cave, New Zealand, and other caves in New Zealand and Australia, is the larval form of a fungus gnat. Hundreds of thousands of glowworms live on the walls and ceilings of Waitomo Cave and illuminate it with a diffused glow. The largest aggregation of them is associated with the cave stream, but small groups are scattered throughout the cave. They can also live on the surface. These insects can control the intensity of the light they produce, and they are capable of glowing 24 hours a day.

Above ground, the same type of glowworm lives in damp shady ravines or along the banks of streams, and emerges at night to feed. In the cave most of them find shelter in holes in or near the ceiling. A glowworm feeds by producing what appears to be a long thread of silk, which bears a series of mucous droplets at regular intervals, giving the appearance of a string of beads. These "fishing lines" vary in length from 2 centimeters to nearly a meter. In darkness, midges and small flies are attracted upward by the glowworm's light and become ensnared in the sticky droplets of mucus. The glowworm then draws up the fishing line and eats the prey. After eating, the glowworm always removes the remains of the meal from its hole and keeps the fishing line clean and in good repair to be used repeatedly for catching further prey.

Some moths spend the winter in caves. One species *Scoliopteryx libatrix*, found in the twilight zone, shows a distinct orientation to light—it keeps the axis of its body parallel to the light rays, with its head farthest from the light source. These moths appear to sparkle when the light of a lamp is turned on them, the light being reflected from beads of water that condense on their bodies.

Wasps and ichneumon flies also use the twilight zone for shelter during the winter. Some wasps apparently spend their whole life cycle near a cave entrance, keeping inside except when they emerge to feed.

THERE IS A LARGER NUMBER of species of cave beetles than of any other cave-dwelling animal group in the United States. There are more than 200 known troglobitic species and subspecies. Even though sheltered by the cave, many beetles remain under stones and near walls and are therefore frequently overlooked by people traveling through caves.

The features of the troglobitic beetles in one cave population are often distinct from those of their relatives dwelling in a cave only a few kilometers away. Thomas Barr has found that the many species of cave beetles in the United States differ widely in form and structure. In the Ohio River basin and the southeastern states, beetles are common in most of the large caves, each of which may have its own species or subspecies that is commonly absent in practically all the other caves.

Cave beetles generally are colored, their colors ranging from reddish brown to black depending on the species. They not only are blind but also commonly have no optic nerve; their appendages are longer than those of comparable surface forms. Many species of cave beetle have powerful jaws, and these species have been seen to feed on millipedes, other beetles, insect larvae, and cave-cricket eggs.

Beetles abound in caves that contain large numbers of bats; they consume the carcasses of bats that litter the floor, and they feed on bat guano. Beetles are preyed on in turn by cave crickets and spiders. Notwithstanding the great abundance of American cave beetles, little is known of their mating habits, egg depositing, or larval development.

CAVE CRICKETS are the most conspicuous inhabitants of large caves, where swarms of them live on ceiling and horizontal surfaces. The bodies of the adults are about 5 centimeters long, but their average length overall, when measured from the tip of the antennae to the end of the hind legs, is about 15 centimeters. A New Zealand species attains a length of 45 centimeters.

Most cave crickets are not troglobites; they migrate to the entrance and feed outside when environmental conditions approach those of a cave—damp, dark, and neither too warm nor too cold. Only one genus of cricket in the United States *Ceuthophilus* is thought to contain some species that are troglobitic. These live in caves in the Rocky Mountain states, and the best-studied one is *Ceuthophilus longipes*, found in Carlsbad Cavern and other caves nearby.

The eastern cave cricket *Hadenoecus subterraneus* roosts near cave entrances and forages outside, mainly as a scavenger. When mature, the crickets migrate deeper into the cave to sandy areas where they mate. The female lays her eggs by plunging her ovipositor into sandy silt and depositing the eggs beneath the surface. The embryonic development takes from 1 to 2 months, and the shorter of these times favors egg survival, because eggs are being steadily searched for by beetles. For the eggs that do hatch, the immature crickets, or nymphs, move to the ceiling to complete their development, thus avoiding predation by the beetles. The nymphs usually roost apart from the adults. This is another means of self preservation, for the adult crickets will eat the nymphs.

In Mammoth Cave, Kentucky, Thomas Poulson has studied the interrelations between the cave cricket *Hadenoecus subterraneaus* and the beetle *Neaphaenops tellkampfi*. The cricket, which feeds outside the cave, supplies food in the form of its eggs to the cave-restricted beetle. The female cricket deposits her eggs at a uniform depth of 13 mm, and she then uses her ovipositor to rake a broad mound of sandy silt over the site. If the blind, 5-mm-long beetle finds the mound, it digs 11-mm-deep holes in search of the egg—a depth just sufficient to reach the top of the 2-mm egg. Each hole, successful or not, takes about 1 hour, and digging for eggs consumes about two-thirds of the total energy expended by the beetle during its lifetime.

When a beetle becomes aware that it is not the only beetle probing a mound, it hastily digs holes that are too shallow to reach the egg.

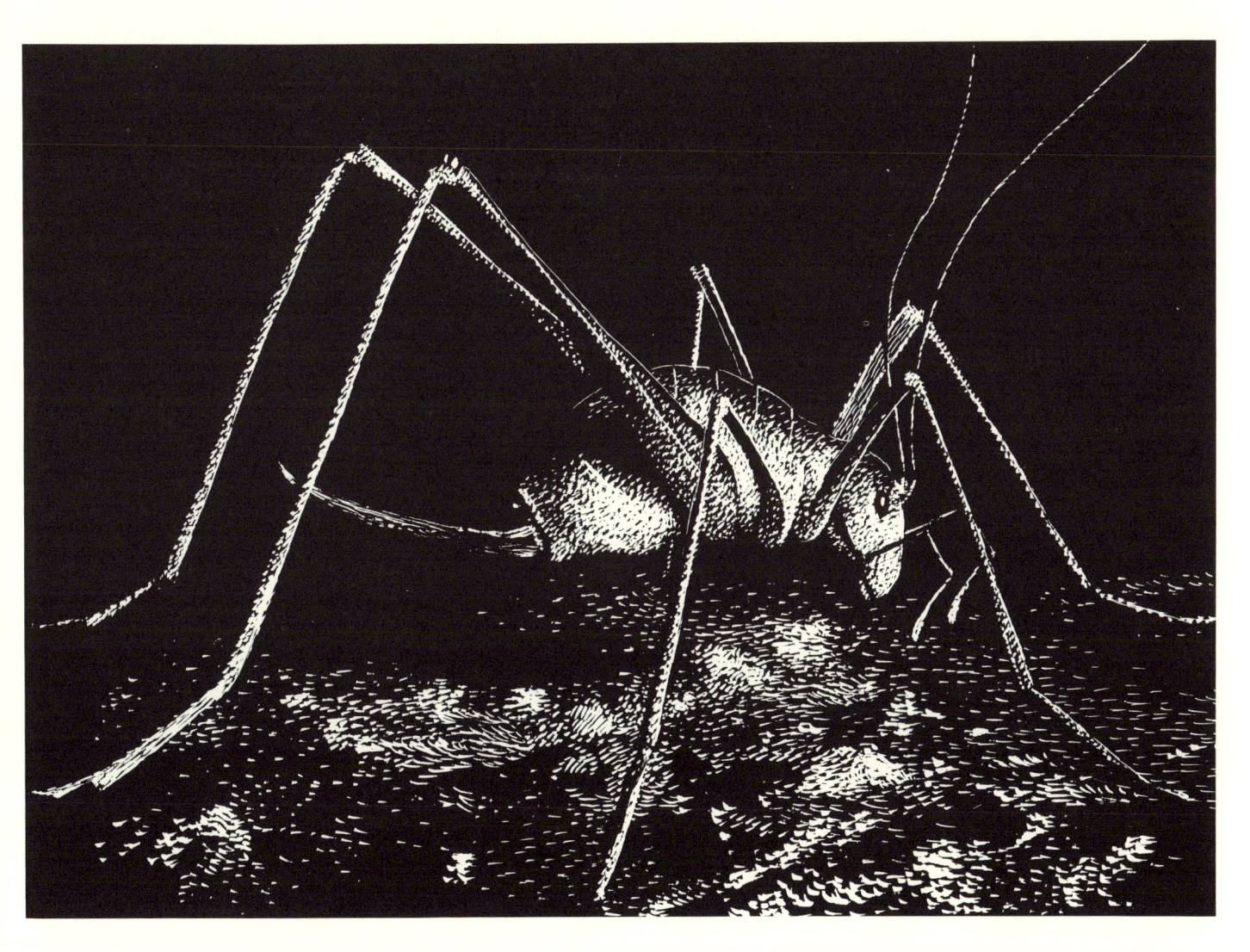

The cave cricket *Hadenoecus subterraneus* does not chirp as its surface relatives do. Its antennae, or feelers, are longer than its body.

102   6 / *Habits of Cave Animals*

Then it tries to drive off the interloper, because fighting takes less energy than digging.

The average beetle gets one cricket egg every 1 to 2 months. It then carries its find under a rock and eats it. When egg mounds are abundant, the beetles find 80 percent of the eggs, but when the mounds are few, the success rate is much less, and the local beetle population declines.

In cave-cricket roosts near cave entrances, cricket droppings adhering to the wall and floor are a source of food for the larvae of beetles and flies and for snails, springtails, and millipedes. The main predators of crickets are large cave spiders whose webs ensnare the crickets as they leap through the air. Salamanders also feed on crickets, and frogs are occasionally observed near a cave entrance, whipping out their tongues to catch them.

The long antennae of cave crickets continually move, sensing air currents that carry odors that could lead them to food. The long antennae also enable them to sense obstacles in the cave darkness. The animals display a homing tendency. Our study of cave crickets in Cathedral Cave, Kentucky, and similar studies in New Zealand, indicate that they always return to the same site in the cave after foraging outside for several hours.

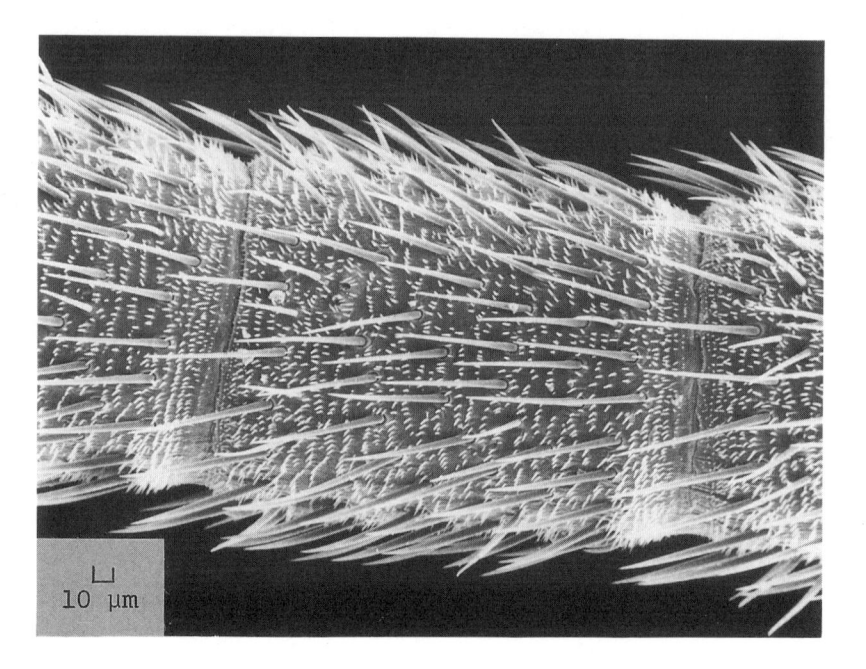

Scanning electron micrograph of the antenna of the troglobitic cave cricket *Ceuthophilus utahensis*, showing hair-like tactile organs. (*Micrograph supplied* by Ruth Simon.)

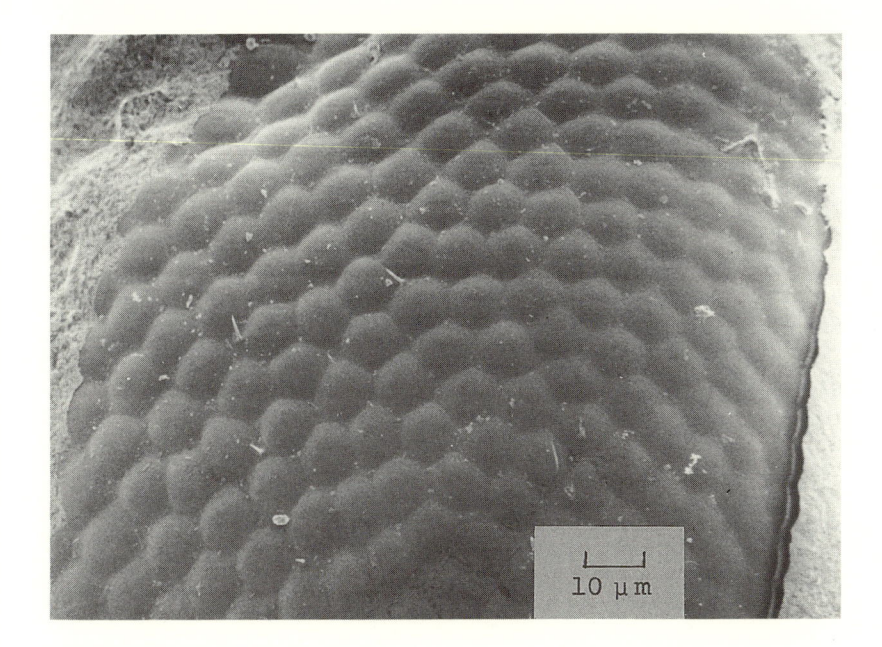

Scanning electron micrograph of the top half of the compound left eye of *Ceuthophilus utahensis*, showing facets assembled in the pattern of a geodesic dome. The eye is smaller than that of its surface relatives, and the facets are fewer in number and less sharply defined. (*Micrograph supplied by* Ruth Simon.)

SEVERAL DOZEN small, white, eyeless species of troglobitic spiders have been found in North America and Central America. The more common cave-dwelling forms, however, such as *Meta menardii*, are large and pigmented trogloxenes or troglophiles. Their habits differ little from those of spiders found in gardens. Their webs trap mosquitoes, flies, beetles, and crickets; their egg cases are usually attached to the cave ceiling. From 20 to 40 young hatch from each egg case of *Meta menardii*.

The cave harvestmen, or opilionids, are sometimes called "daddy longlegs." They have circular bodies about 4 millimeters in diameter, from which extend eight sharply jointed legs. *Phalangodes armata*, a troglobitic species found in the caves of Kentucky and Tennessee, is known to feed on springtails, but much study is needed to determine the habits and life cycles of these spiderlike organisms. Swarms of trogloxenic harvestmen are readily detected by a pungent odor that emanates from them.

Many mites found in caves are parasitic on bats. Ticks may be found in the twilight zone, where they wait for a suitable host to latch onto, such as a raccoon or a skunk that enters the cave for shelter. After filling themselves with blood from their host, the ticks drop off. When their meal is digested, they are once again on the alert for a victim.

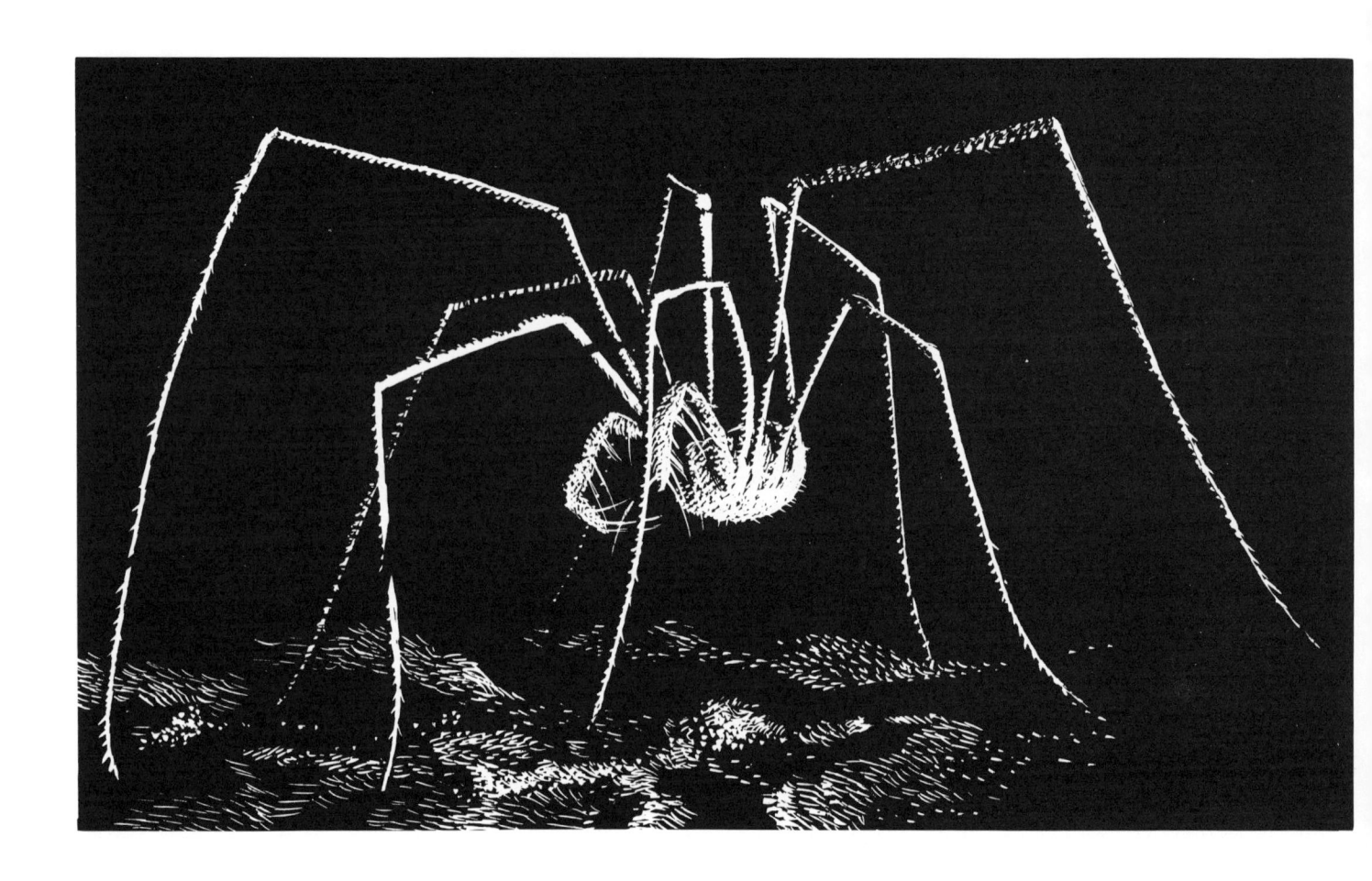

The blind harvestman *Phalangodes armata* continually waves its long front legs as it walks over the walls of a cave.

Vertebrate Cave Animals

THREE FORMS OF VERTEBRATES are commonly found in caves—fish, salamanders, and bats. Troglobitic species are known for the first two. Cave bats are all trogloxenes.

The first cave fish discovered in the United States *Amblyopsis spelaea* was described in 1842. It is found in the caves of Kentucky and southern Indiana. Because of its unusual degeneration of pigment and eyes, it was intensively collected during the 19th Century and sold as a souvenir, with the result that it has become locally rare. Another species, *Typhlichthys subterraneus*, also in the family Amblyopsidae, has a wide range in the states of Kentucky, Tennessee, Missouri, and Alabama. Two other species of troglobitic fish also occur in the central United States, and two species are known from Texas.

Cave fish are rarely more than 10 centimeters long. Sensory projections scattered over their bodies allow them to detect vibrations of small animals in the water and enable the fish to obtain food in total darkness. These same receptors are used by the fish to detect obstacles that are being approached, and permit the fish to avoid them. Because of their ability to detect vibrations, they are difficult to catch. Slipping a net or even one's hand into water a meter from a fish may cause it to dart away. They tend, moreover, to remain close to the banks of stream passages, where they are hidden by overhanging rocks and ledges.

The female's oviduct is in front of the gills, directly behind the mouth, and the fertilized eggs are retained in the mouth cavity. This affords protection for the young until they are able to forage for themselves. There are no known predators on adults. Cave fish feed on copepods, isopods, amphipods, and small crayfish. They are also known to be cannibalistic.

A study of cave fish in the United States shows that they can move from one cave to another through subterranean channels below the water table. *Typhichthys subterraneus* is found in Tennessee on opposite sides of the Cumberland River. Although the river might seem to constitute a barrier to cave fish, continuous beds of limestone pass beneath it. Cave fish are known to live in cave openings in the bottom of the river. Hence it is possible that they once migrated in water-filled dissolution channels in limestone below rivers of larger size and thicker alluvial deposits, such as the Mississippi River.

The Kentucky cave fish *Amblyopsis spelaea* has degenerate eyes. Vibration receptors, which are scattered over its skin, allow it to detect the faintest vibrations and guide the fish to swimming organisms that serve as its food.

SALAMANDERS were the first cave vertebrates to be studied carefully. The first one to be described, in 1768, was *Proteus anguinus*, found in Slovenian caves. But even though the caves of Europe have been intensively explored for more than 200 years, no other troglobitic salamanders have been found there. In the United States, however, nine species of troglobitic salamanders are known: One on the Ozark Plateau, five on the Edwards Plateau of Texas, and three in the cave areas of the southeastern states. Many other species, including the so-called cave salamander *Eurycea lucifuga,* are trogloxenes.

The troglobitic forms have much reduced pigment and eyes, whereas the trogloxenes are well pigmented and have normal eyes. Most troglobitic salamanders are neotenic forms—that is, they have never progressed further morphologically than the larval stage. Physiologically, however, they are mature breeding adults when full grown.

Most troglobitic salamanders breathe through their skin, which is directly underlain by a dense network of capillaries. This permits the exchange of carbon dioxide and oxygen as long as the skin remains moist. These blood vessels give the skin a purplish tint that makes the animals seem pigmented. All cave salamanders feed on cave flatworms, small crustaceans, and insects, but trogloxenic forms also feed on surface organisms near the moist rocks and crevices of the entrance.

The Texas cave salamander *Typhlomolge rathbuni*, which lives in some of the water-filled caves that constitute the Edwards Limestone aquifer in central Texas, has figured in local water-conservation efforts. Texas water law had followed the so-called "capture" principle, which permitted everyone to pump whatever water they wished from that available underground. Plunging water levels in the closely integrated cave water system became alarming. But the cave salamander is an officially listed endangered species, so lacking a direct legal mandate to conserve water, those who faced the impending water crisis initially used the endangered-species law to protect the habitat of the salamander. By so doing, they preserved the economy of the city of San Antonio and its surrounding region.

Most local people did not see the protection of an endangered species as more important than the short-term interests of humans, however, and the topic fueled a sentiment that "people are more important than critters." But the cave-salamander issue has bought time, and the citizens now slowly seem to be recognizing that the economic prosperity of the region depends on the sustainability of the local water resources.

The troglobitic salamander *Typhlomolge rathbuni*, which lives in submerged caves in Texas, breathes through red external gills. Other species of blind cave salamanders have no gills and breathe through their skin.

SNAKES do not ordinarily crawl into the dark zone of caves because they dislike its excessive moisture. Occasionally, however, one will accidentally drop into a sinkhole that leads into a cave. Caves in Malaysia contain a species of bat-eating snake that is pale but has normally developed eyes. This reptile crawls through the cave, quickly wraps itself about any sleeping bat it finds, crushes the bat, and swallows it headfirst. Snakes frequent the entrances of large bat caves in the southwestern United States and feed on young and injured bats.

Snakes that do not live permanently in caves often use the entrances as dens. Some of the largest rattlesnake dens in Texas and Oklahoma are located in dry cave entrances that maintain a fairly even temperature. Copperheads are also occasionally seen just inside entrances to large caves. Cave explorers have been bitten by them while climbing onto ledges in the twilight zone.

THE CAVE RAT *Neotoma magister*, which is related to the familiar pack rat, has the same habit of accumulating in its nest all sorts of shiny things, from candy wrappers to flash bulbs. Cave rats, like pack rats, differ from house rats in having furry tails and white hair on the abdomen. Extremely inquisitive, cave rats show little fear of people. They leave the cave for grain and seeds, the main constituents of their diet, but also feed on such other kinds of plant material as fruit, leaves, and roots.

Cave rats have large eyes, but in the darkness of the cave they feel their way by means of vibrassae—long white whiskers on the face. And, as we note above, when deep within the zone of total darkness they leave trails of urine droplets, which they afterward follow by scent. Cave rats are sometimes found inside a cave several kilometers from the nearest known entrance. It is possible, however, that they can return to the surface through crevices too small to be entered by people.

The droppings of cave rats are eaten by larvae of beetles, and they are also, indirectly, a source of nourishment for insects that feed on the fungi growing on these droppings. In northern latitudes, the rats store up enough food in their nests to carry them through the winter.

In desert caves of western North America, related species have left deposits that in places are more than 40,000 years old. The older parts of these deposits commonly contain well-preserved twigs from coniferous trees that date from the cooler and wetter Pleistocene Epoch, and which are now absent from the area of the caves.

Vampire bats sip up blood after lightly cutting the skin of a victim with slightly protruding incisor teeth. The bat's nose is blunt, and its teeth have such sharp edges that they make a clean incision, often without waking the victim.

BATS are the only true flying mammals. Their wings are membranes extending from the bones of the forearm to the rear of the body. Fur covers most of the body, often obscuring the eyes.

In the darkness of a cave, a bat navigates by a method called *echo location*, using its voice and ears. Its outer ear is large, having a broad scooplike form, and projecting well above the head, so it easily gathers sound waves. A bat's sense of hearing is amazingly keen. While flying in caves it continuously emits ultrasonic squeaks that are inaudible to humans, but are audible to the bat, not only directly but also when echoed from objects a few meters away. A bat can judge its distance from the echoing object by the length of time it takes the echo to return, and it can estimate the approximate size of the object. If it is large, the bat steers clear of it; if it is small and in motion, the bat seemingly infers that it may be an insect and dives quickly toward it. The bat scoops up its prey in the membrane of its wings or tail, then doubles over to grasp the insect with its strong jaws.

Bats are not blind. When their ears are plugged so they can no longer determine their position by locating echoes, some species can, if enough light is available, avoid obstacles and catch food by using their eyes.

Cave bats fly about outside the cave at night, and the objects they usually dive at—and usually catch—are insects. A bat commonly eats, on the average, about half its weight in insects every day that it is not hibernating. The 100 million Mexican free-tailed bats living in the caves of the southeastern United States and Central America eat an estimated 100,000 tons or more of insects each year. This immense amount of food is the source of the large quantity of guano that accumulates in bat caves. A few of these bat caves have been mined for guano, which is rich in nitrogenous material and hence is an excellent fertilizer. Layers of guano 12 meters thick have been found in rooms of caves in Carlsbad Caverns National Park and in caves in the Grand Canyon.

Aside from its economic value, guano is also important as a source of food for other cave-dwelling animals. Cave explorers may find crawling into a guano cave unpleasant because of the guano's mushy consistency and fetid ammoniac odor, and also because of the innumerable beetles, ticks, lice, and mites that swarm over it. To those creatures, however, the guano is welcome food. The heaviest concentration of cave life is where guano abounds.

For more than 50 years, naturalists have been investigating the movements of bats by banding them with lightweight clips. The bands, usually placed around the equivalent of the forearm, do not interfere with flying. To date, half a million North American bats have been banded. If you find a banded dead bat, you should send the information on the band to the United States Fish and Wildlife Service, Washington, DC 20240.

Studies of bat-banding records have established that some bats live for 20 years. These same studies have disclosed that bats return to the same cave year after year, although they may migrate as far as 1,200 kilometers from the cave during the summer. Experimenters have demonstrated bats' homing instinct by trapping them in a cave, enclosing them in a covered cage, and transporting them to a site 800 kilometers from the cave. When released, the bats circled about for several minutes, then seemed to disappear into the sky. Within three days, 90 percent of the bats had returned to the cave.

Bats usually mate in the latter part of the fall. In some species, fertilization occurs immediately, but the embryonic development is latent during the winter. In other species, the spermatozoa remain active in the oviducts until spring, when fertilization finally occurs.

In all species, the young are born in late spring and early summer, and the normal number of young is one. The infant usually appears feet first, while the mother hangs by her hind feet or her thumbs, or both. She curls the tail membrane around her belly as a sort of pouch to receive the newborn bat, which immediately attaches itself to its mother's breast and begins to feed. For a week or so the young bat rides along with its mother as she flies, keeping a tight hold by biting into her fur and digging in with its claws. Later most of the young bats can hang on the walls by themselves while the adults are feeding, but some of them, being too weak to maintain their hold, fall to the floor, where they are attacked by beetles and mites. Mother bats do return to their own babies. After a month or so, young bats can fly and feed on insects.

Bats, when resting, hang head downward. On each hind limb are five toes, and on each toe is a sharp-hooked claw that enables the bat to keep its hold on the slightest surface irregularity. Although bats can crawl, their movement is awkward and shuffling, and they are easily captured by predators on the ground. Skunks, raccoons, foxes, snakes, and ringtails capture bats that have fallen near cave entrances. Outside, hawks and owls occasionally catch flying bats on the wing.

The bat's normal body temperature, like that of a human being, is about 37°C. Northern species that do not fly south in the fall must hibernate because few insects are available during the winter. Some species, such as *Myotis sodalis* in Indiana, huddle in clusters when hiber-

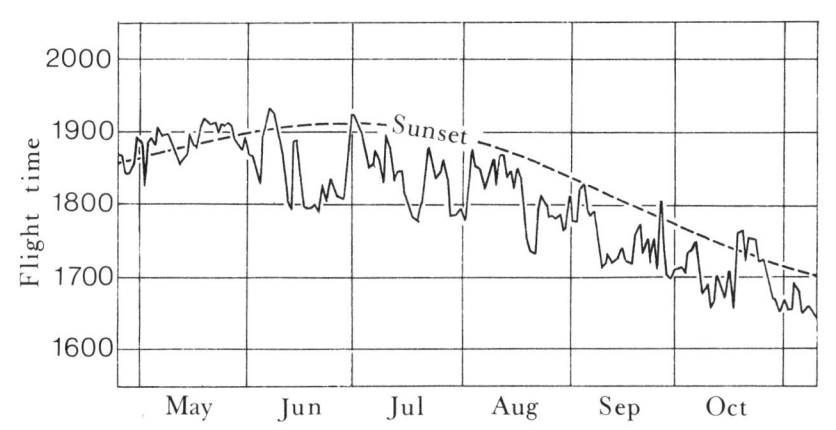

Times of the evening departure of the bats from Carlsbad Cavern. The bats usually leave immediately before sunset. We suggest that their timing is based on the sudden rise in barometric pressure that normally accompanies sunset. If this hypothesis is correct, variations in flight time could result from fluctuations of the normal pressure caused by local weather conditions. (*These unpublished data, for 1942, were supplied by* James Baker, U.S. National Park Service.)

nating. Others, like the so-called solitary bats, hibernate singly. During the period of hibernation, the body functions are so suppressed that they are barely perceptible. The bat's heartbeat and blood flow are almost quiescent; it may take a breath no oftener than once every 3 minutes; its body temperature drops as much as 30°C below normal. Digestion ceases; the hibernating bat gets energy from fat stored up in the autumn. By the spring the bat has lost as much as a third of its weight and is emaciated.

Hibernating bats apparently do not sleep continuously but occasionally move about, probably to drink, because water constantly transpires from the thin, naked wings, ears, and tails. Observations on marked specimens indicate a shifting of position every week or so, and even an occasional short flight outside the cave when the weather is not too severe.

BIRDS RARELY INHABIT DEEP CAVES, although phoebes and cliff swallows commonly build nests inside cave entrances. In Texas and Mexico, some cave colonies of swallows have thousands of members.

The famous bird-nest soup, greatly relished by affluent gourmets, is made from the nests of a cave-dwelling swift that lives in southeast Asia. The cup-shaped nests, which are attached to the limestone walls of the caves, are built from coagulated saliva secreted by the birds. Collecting the nests is a thriving local industry for people in Borneo and Thailand.

Nest collecting has reduced the bird population and driven them to the highest ceiling levels. Families that collect the nests now carefully regulate the harvest to avoid reducing the population further, but poaching by outsiders is at times a problem.

Collecting is done from a bamboo scaffold attached to the ceiling as much as 20 meters above the cave floor. Access to the scaffold is by a dangling pole that, to deter poachers, ends about 3 meters above the floor. Collectors arriving for work carry a collapsed bamboo tripod to gain access to the pole.

Two grades of nests are produced for the market. Experience has shown that the birds will build a second nest if the first is collected before they lay their eggs. The first cream-colored nest constitutes top quality and commands a high price. After it is collected, the birds are allowed to build again and to raise their young. The second discolored nest is then collected and carefully cleaned, and it is marketed separately.

A bird that lives much of its life in the dark zone of caves is the guacharo or oilbird *Steatornis caripensis*, which is found in caves of Venezuela and Trinidad. It avoids obstacles while flying in the dark by a method of echo location akin to that of bats, except that this bird produces clicks audible to humans. The guacharos leave the cave at twilight to feed on the fruit of forest trees. They may fly as far as 30 kilometers from the entrance before returning at dawn. The floors of the caves occupied by guacharos are littered with seedlings that have started to sprout from seed passed in the birds' droppings, but these seedlings soon die for want of light.

Guacharos spend most of the daytime in pairs by their nests in the cave. These nests are extremely simple, and the eggs are laid on the bare rock of ledges close to the ceiling. As might be expected, many of the eggs roll off and fall to the floor, which at the climax of the egg-laying season resembles a giant omelet. The young are fed at intervals during the night with food regurgitated by the parents. Young guacharos are very fat, so they are able to keep warm in the depths of the cave even before they acquire their downy feathers.

Segregation of Animals

THE ORGANISMS PRESENT IN A CAVE do not intermingle at random, but are usually more or less concentrated in specific areas. Aquatic forms, such as flatworms, amphipods, isopods, and crayfish, normally can live only in water, although in caves where the humidity approaches saturation, the smaller aquatic animals may crawl over damp rock. Worms, beetles, millipedes, and springtails live in or near mud. Cave crickets, spiders, and harvestmen congregate on the walls and ceiling. Bats ordinarily hang from the ceiling or from crevices close to it.

This segregation within caves consists of two types: stratification and zonation. Stratification results from the gathering of individual species in localized areas on the floor, the walls, or the ceiling. Zonation derives from the preference shown by given species for the entrance zone, the twilight zone, or the dark zone of a cave. The three zones and the stratification of some typical inhabitants of each are shown in the accompanying table.

Stratification and Zonation of Cave Animals

	Entrance zone	Twilight zone	Dark zone
Ceiling	Mosquitoes Moths Wasps	Crane flies Crickets Spiders	Bats Crickets Spiders
Walls	Centipedes Salamanders Millipedes	Harvestmen Salamanders Millipedes	Harvestmen Salamanders Millipedes
Floor	Cave rats Snails Springtails	Beetles Snails Springtails	Beetles Crayfish Fish

Cave animals also segregate according to subcommunity type based on the kind of available food, such as cave-rat droppings, cricket guano, or leaf litter. A few organisms move across the boundaries of the various zones, particularly if they are predators or if they are leaving the cave to feed. Our experiments with marked crickets in Mammoth Cave National Park indicate, however, that after feeding, the great majority of the animals return to exactly the same place in the cave from which they departed in search of food. Because some of these areas are perpetually dark, the cave dwellers must have some olfactory faculty or some other sensory receptor that enables them to return repeatedly to the same resting place. The failure of marked crickets to return to areas accidentally wetted with alcohol provides evidence that olfactory sensation may play some part in this homing instinct. Even after the alcohol had evaporated, returning crickets seemed to be confused on reaching the site where they normally would have rested.

Cyclic Behavior Patterns

BECAUSE MANY CAVE ANIMALS live in an environment of total darkness and uniform temperature, you might suppose that they would display no rhythmic seasonal or daily changes in their activity. Most of the smaller troglobites, indeed, seem to be devoid of such rhythms. Some larger forms, however, such as crickets and crayfish, do have definite cycles of activity. Some species of cave crickets, including those that stay in the area of total darkness during the day, move out of the cave at twilight to feed and return before dawn. Bats likewise leave the cave daily when the outside temperature is above 7°C. Crickets will forage outside the cave in temperatures as low as 2°C, if the relative humidity outside the cave remains above 85 percent.

The eyeless cave crayfish *Orconectes pellucidus* can display an approximate daily cycle of activity, even though the species has not been exposed to daily light and temperature changes for thousands of generations. Mammoth Cave crayfish held under simulated cave conditions in a laboratory walked and foraged for food more than twice as actively at 7:00 p.m. as at 10:00 a.m.

Pioneer studies on several phyla of troglobites show that these animals have a clear circadian (nearly daily) rhythm. Recent studies in the United States of the activity of cave crickets of the genus *Ceuthophilus* and in France of cave millipedes of the genus *Blaniulus* have shown that they have, in addition to the circadian cycle of 24-26 hours, a secondary cycle of 12-13 hours. This secondary cycle is in phase with the earth tides, of which two cycles take place every day.

Two tidal cycles occur because every point on the spinning Earth passes by two regions of low gravity each day. The Moon's gravity distorts the Earth into an elongated ellipsoid whose long axis points toward the Moon. Points on the Moon-facing and Moon-opposite bulges of the ellipsoid lie farther from the Earth's center of mass; hence they have lower gravity than intermediate points. Cave animals (and noncave animals as well) may key some of their biologic rhythms not only to a sensing of effects related simply to the once-daily rotation of the Earth, but also to the pervasive twice-daily changes in gravity related to the tides.

More work is needed on other troglobites to determine whether or not they also show twice-daily peaks of activity. Whatever the cause, the response will probably be found to be similar to that of other animals. Even the life processes of incubating chicken eggs seem to be controlled by biologic clocks.

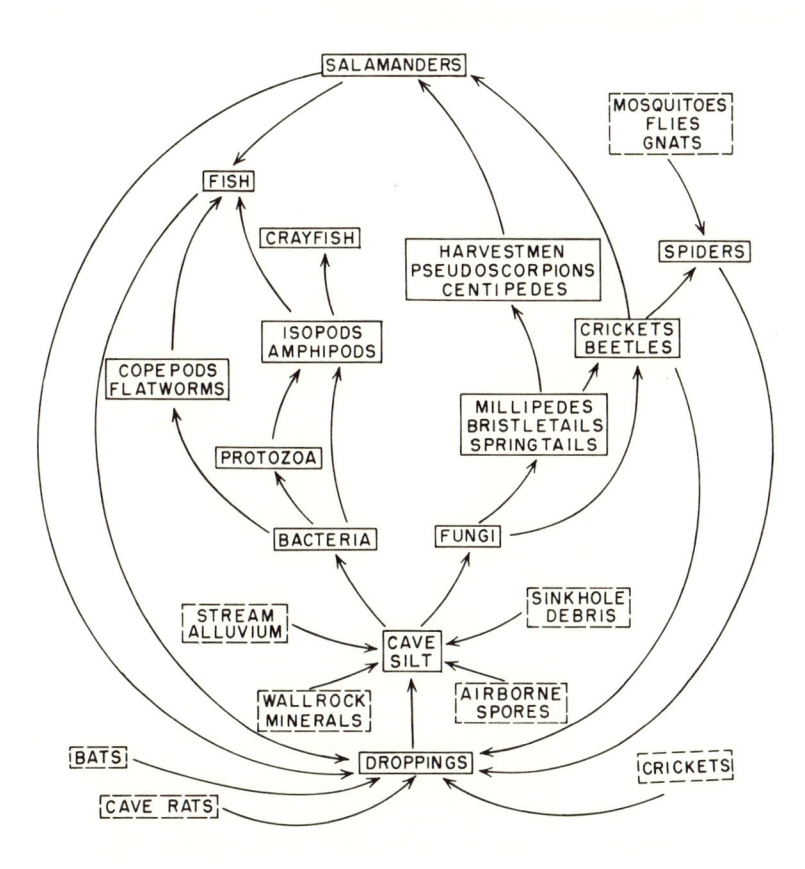

The cave food web. Arrows indicate the direction of energy flow supplied by food. Components in dashed boxes are derived from outside the cave.

Ample evidence exists for annual periodicity of reproduction in many cave organisms. For trogloxenes, which periodically leave their caves, this is clearly related to the surface influence. For troglobites, it may relate to seasonal changes in food availability, such as that brought in by the trogloxenes. In cave crayfish, a nearly annual clock governs both reproduction and molting. Under constant conditions with no light in a laboratory, this cycle persisted for two years.

Many species of cave salamander, being similar to closely related surface forms, have a specific breeding season, which results in egg-laying periodicity and hatching of the young at specific times of the year. The bats of northern North America are well known for their already described habit of either hibernating or migrating south to warmer climates in the winter.

Cave Food Web

THE LACK OF LIGHT precludes the existence of most green plants in caves, because the Sun's rays are necessary to provide the energy required by chlorophyll to synthesize the carbohydrates and amino acids needed for plant growth. Seedlings that are occasionally washed into a cave, or carried in by animals, soon wither and die.

Because no photosynthesis occurs in a cave, most nutrient material must be brought in from the surface. This material may be carried directly to the interior in flowing water and sinkhole debris, or indirectly in the droppings from animals that feed outside but return to the cave for rest. The food supply available to troglobites is limited, however, and would not suffice without extremely efficient utilization of organic material. The continuous recycling of material among the cave populations is known as the *cave food web*.

Spores of fungi are carried by slight air currents and on the bodies of animals to many parts of the cave. These spores are particularly abundant in caves that contain large quantities of decaying organic material, or of mud with a high percentage of humus. Fungi obtain nourishment by breaking down and absorbing the organic substances in this material and in animal droppings. Fungus-eating insects, such as beetles, springtails, and mites, feed on molds and bacteria. Cave animals that live in water may also ingest floating organic material directly. All of these animals, in turn, serve as food for larger predators such as salamanders, crayfish, and fish. Organic material also is returned to the cave environment from the dead bodies and droppings of the larger animals; in some caves, bat guano is an important contributor to the reservoir of available nutrient material.

Francis Howarth has demonstrated the essential role of roots and vines in maintaining the fauna of lava caves in the Hawaiian Islands. Although these caves are not deep, they contain a surprisingly abundant troglobitic fauna.

An interesting example of the importance of organic products from outside limestone caves to the food chains within them was discovered by John Holsinger in Banners Corner Cave, Virginia. Sewage from a nearby village, draining into the cave, caused a dramatic increase in the population of certain cave animals. Isopods of the genus *Asellus* were nearly twice as numerous as in most caves, and planarians and earthworms were also unusually abundant. Apparently the cave animals are feeding on a bacterial scum that grows on the surface of the sewage. This particular cave food web had been simplified by pollution, with some of the usual cave species becoming extremely abundant at the expense of other species. Now the polluting has stopped, but the return to the original community structure is taking place very slowly.

The cycling of food in caves often approaches what is known as a *closed ecosystem*. In a completely closed system, every organism feeds on certain other organisms, and in turn is eventually fed upon by still other organisms within the system. But such a system ordinarily cannot maintain itself without some indirect help from sunlight. There is never complete efficiency in passing energy from one level of a food chain to another. Energy is lost as heat, and therefore photosynthesized material from the surface, or energy-containing minerals from the cave wall rock, must be utilized to maintain the system. The cave environment, however, displays a higher degree of efficiency than almost any other; this must be so when animals flourish in total darkness many kilometers from the nearest connection with the surface.

Cave Ecology Based on Hydrogen Sulfide

ALTHOUGH MANY CAVE ORGANISMS depend on particulate plant material brought in through sinkholes and flooding, and others depend on the droppings of animals that migrate outside to feed, some cave ecosystems are based on hydrogen sulfide gas that rises from deeper strata. A prime example is Movile Cave, Romania.

Movile Cave was discovered in 1986 when a shaft was dug as part of a foundation study for a never-built power plant. The shaft is near the rim of a sinkhole about 1 kilometer from the Black Sea, and the cave has no natural entrance. Upper passages are dry, and lower passages near sea level are partly submerged. A strong smell of hydrogen sulfide pervades the air near pools that connect with the submerged passages, and Serban Sarbu reports a unique and rich troglobitic fauna in the water and on exposed cave surfaces.

Divers who ventured into the submerged passages found more unusual organisms. The richest communities are in air-filled domes that have no outlet to the remainder of the cave except through the submerged passages.

A mat of fungi and bacteria floats on the water in the air-filled domes, and similar microbial mats cover nearby limestone walls in the domes. The air of the domes contains 2 percent carbon dioxide and about a third of the oxygen of normal air.

The floating mats consist of densely intergrown Oomycetes water-mold filaments and *Beggiatoa* bacterial filaments. They also contain tiny blebs of sulfur. The oxygen-tolerant *Beggiatoa* bacteria in the mats, along with oxygen-intolerant rod-shaped bacteria in the underlying water column—probably *Thiobacillus*—are chemoautotrophic. They use the energy from hydrogen sulfide to make organic molecules by fixing the carbon from carbon dioxide in the cave water in reactions such as the following:

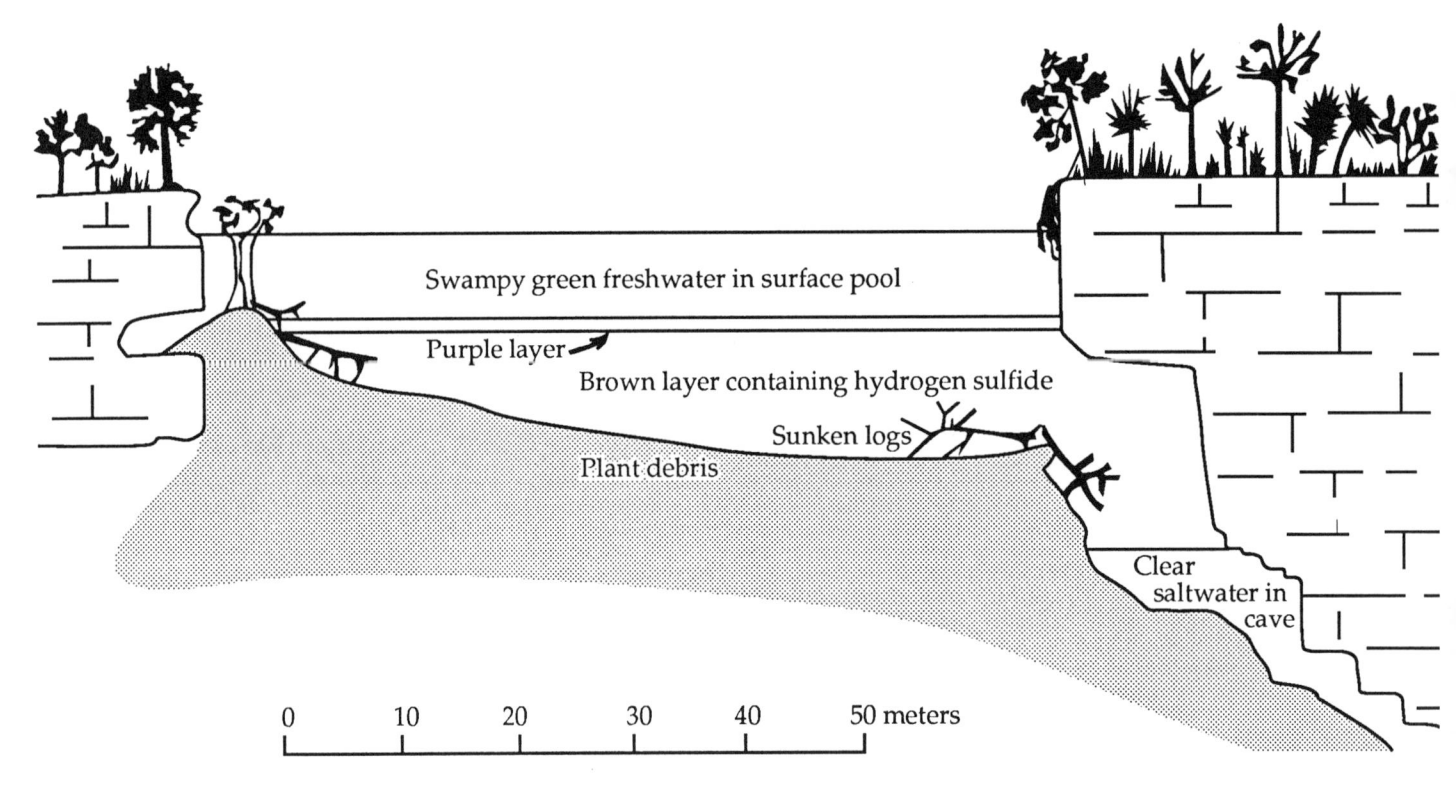

$$\underset{\text{sulfide}}{\underset{\text{hydrogen}}{6H_2S}} + \underset{\text{dioxide}}{\underset{\text{carbon}}{6CO_2}} + \underset{\text{water}}{6H_2O} + \underset{\text{oxygen}}{6O_2} \rightarrow \underset{\text{carbohydrate}}{C_6H_{12}O_6} + \underset{\text{acid}}{\underset{\text{sulfuric}}{6H_2SO_4}}$$

Cross section of Cenote Verde, Mexico, showing fresh- and saltwater layers in the pool and submerged cave below. The upper fresh layer, with a September 1990 temperature of 29.7°C, contains surface catfish and mosquito fish; the purple layer, which blocks out light below, contains sulfur bacteria; the brown layer, rich in hydrogen sulfide, contains swimming phantom-midge larva and draped white sulfur bacteria on solid surfaces; the clear saltwater in the cave, with a temperature of 24.9°C, contains blind cave fish and shrimp. (*After* William Wilson and Thomas Morris.)

The chemoautotrophic bacteria in turn supply organic molecules to feed other bacteria, which are heterotrophic, and also the fungi in the mats. The mats then support a rich and diverse cave fauna.

Protozoans, rotifers, copepods, amphipods, isopods, snails, flatworms, water scorpions, and leaches live near the surface of the water at gaps in the mats where they obtain free oxygen. Springtails, isopods, and millipedes graze on top of the mats. Beetles, spiders, pseudoscorpions, and centipedes are predators that live on and near the mats and pools. In all, 48 species in Movile Cave are troglobitic.

The ecosystem at Movile Cave resembles that of black smokers—hot deep-sea vents along seafloor spreading axes—where a rich fauna of tube worms and other animals also has its basis in chemoautotrophic bacteria that derive energy from upwelling hydrogen sulfide. Great caves such as Carlsbad Cavern and Lechuguilla Cave in New Mexico, which were formed by sulfuric acid derived from hydrogen sulfide, probably had a biota similar to that of Movile Cave during their periods of active growth.

Troglobites From the Sea

DURING THE PAST 20 YEARS, speleologists have discovered many saltwater troglobitic organisms in tropical caves that connect inland surface pools with the sea. L.B. Holthuis proposed the term *anchialine* for the saltwater caves containing these organisms. Some far-inland and ordinary freshwater aquatic troglobites, such as isopods and amphipods, may have had an early history as coastal troglobites in such seawater-filled caves. Over the millennia, they gradually adapted to freshwater and joined the terrestrial troglobites that originated from the land surface.

Diving speleologists have found unusual conditions in the generally circular pools that serve as windows into tropical saltwater caves. A lush vegetation surrounds the pools, and floating algae make the upper layers of the water look like pea soup. These layers are generally of freshwater that floats on the saltwater of the cave with a sharp discontinuity at depths ranging from 10 to 20 meters. Many of the saltwater caves contain submerged stalagmites that indicate a former air-filled history before sea level rose to its present level about 5,000 years ago.

In Cenote Verde, a karst pool in southeastern Mexico, William Wilson and Thomas Morris found a remarkable sequence as they descended below a swampy green layer at the surface of the pool. The freshwater upper level contains surface catfish and mosquito fish. With increasing depth, the freshwater changes to brackish, and its green color changes to a nearly opaque brown. Hydrogen sulfide gas pervades the water at this level, up to more than 25 milligrams per liter, at which concentration it burns the exposed skin of the divers and blackens the chromium-plated parts of their diving gear. Surprisingly, this seemingly inhospitable layer contains living phantom-midge larva.

At the top of the brown zone is a thin purple layer containing sulfur bacteria. Thick coatings of white filamentous sulfur-reducing bacteria also coat the rock walls and tree limbs that have fallen into the pool. Below that lies crystal-clear seawater that is well oxygenated and contains troglobitic blind cave fish and shrimp.

These chemical conditions in Cenote Verde and similar karst pools in Mexico, Cuba, and the Bahamas seem to result when organic matter in and around the pools dies and chemically reduces sulfate ions from the underlying seawater to hydrogen sulfide.

CERTAIN PHASES of the natural history and life cycles of cave animals are not yet understood. The prime requisite for filling these gaps in our knowledge is the continued observation of these organisms in their natural habitat. Nothing should ever be done, therefore, to disturb or harm cave life greatly or irreversibly. Only biologists trained to discriminate new species, or those engaged in physiological studies of cave life, are justified in removing organisms from a cave.

The eyeless nonpigmented European salamander *Proteus anguinus* swimming in a cave with its young. The regressive characters appear late in the growth of this species; the young have eyes and small color spots.

7: Evolution of Blind Cave Animals

MOST OF THE ANIMALS that live in any cave today are descended from ancestors that have lived in the same cave for many thousands of years. Hence it is not remarkable that species restricted to caves have characteristics, such as blindness, that distinguish them greatly from their surface relatives. Since normal hereditary changes have affected the permanent cave dwellers, or troglobites, for thousands of generations, the wonder is that these animals do not differ even more from their surface-dwelling ancestors.

The great majority of all troglobites alive in North America today are descended from species that originated during or before the last glacial stage of the Pleistocene ice age, which came to a close over a period of several thousand years beginning about 15,000 years ago. When we plot the present distribution pattern of troglobites in both the Eastern and Western hemispheres, we see plainly that few troglobites now live in caves that were covered by ice during the last stage of the Pleistocene Epoch. Life existing in caves within the limit of this last glaciation was largely exterminated by the thick cover of ice, and not enough time has passed since then for new troglobites to evolve.

Colonization of Caves

WHY AND HOW did surface forms come to populate caves? Actually, any species will tend to migrate into, and fully populate, any environment in which it can survive and reproduce. Once it becomes established, qualities that help it to survive become highly developed, and those that are detrimental waste away. However, the organisms that have migrated into caves were fairly well adapted beforehand for life in this environment. They have been, in a word, preadapted.

Relationship between the ranges of troglobitic fish species and the southern edge of Pleistocene glaciers. Insufficient time has elapsed for the evolution of new cave-adapted species in the area covered by the glaciers since they retreated.

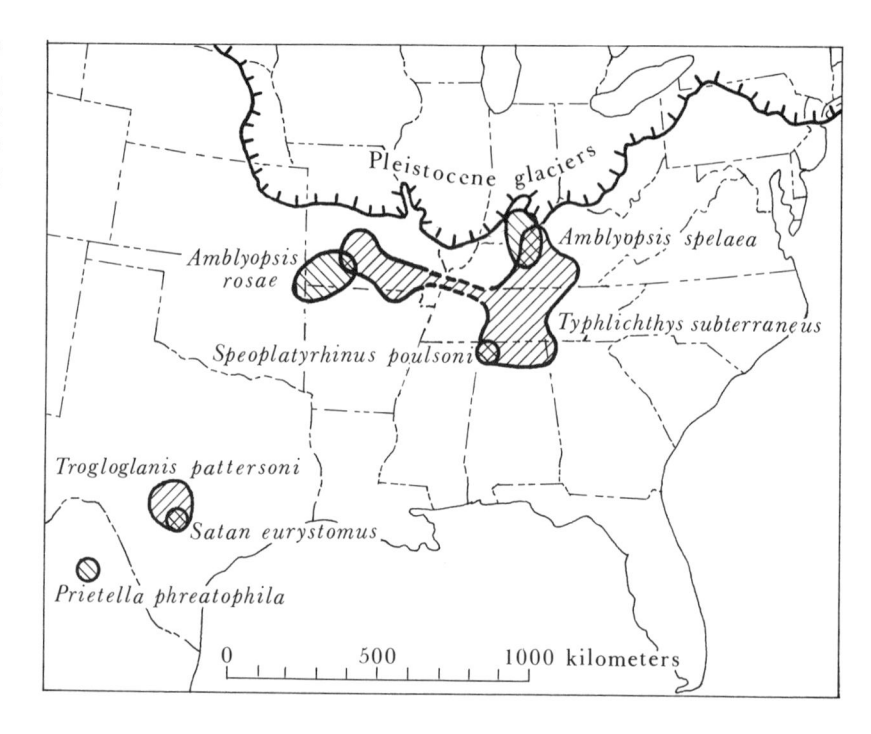

Conditions and events in the past that have been responsible for the initial colonization of caves were not the same for all species. Some organisms entered the caves through wide openings, but most have found access to them through narrow crevices and holes in the surrounding rock. Such animals as millipedes, beetles, snails, and worms normally enter cracks and crevices, and it is such forms as these that make up the dominant part, numerically, of cave populations. These creatures hide in holes, and even in the pitch-dark and humid atmosphere of caves they retain the instinctive habit of crawling under stones and into cracks. They are the troglophiles, which love a cavelike environment.

Aquatic organisms are often swept into caves, and some of them manage to survive. For the most part, however, aquatic cave dwellers develop from surface types living habitually in such dim places as hollows under stones or at the bottoms of streams. Study of the cave fish *Chologaster agassizi*, in particular, suggests this mode of entry. Some populations of this species live at spring outlets in swamps at the base of limestone bluffs in southern Illinois. Others live at ground-water level in Mammoth Cave National Park, in caves that are inundated when the Green River—a surface stream—is in flood. Here the populations living within the cave have descended from populations that formerly lived in springs along the Green River and were carried underground by floodwater.

A species may be drawn into a cave by a need for high humidity. Isopods that live under stones or under tree bark require dampness. In dry air they become restless, lose weight, and soon die. Search for a moist environment would naturally lead them to the entrances of caves, and from there to cave interiors.

An animal already accustomed to dim light is especially likely to find darkness tolerable. The European cave salamander *Proteus anguinus* is exceedingly sensitive to light and avoids it. The mud puppy *Necturus,* the nearest relative to *Proteus,* dwells on the bottom of streams of the eastern United States. It lives in dimly lit water, under rock ledges, or in vegetation. Its habits, which are undoubtedly similar to those of the ancestors of *Proteus* when they lived above ground, would account for the species' penetration into caves.

In Hawaii, Francis Howarth has found that troglobitic insects have been able to disperse from old lava tubes into nearby new ones within a period of only 10 years after the new tubes were formed. Because some preadaptation to cave life is necessary, normally no sudden change attends the colonization of caves. Rather, it is a slow process and one that is taking place today.

Although most cave animals were killed by the ice-age glaciers of North America and Europe, some aquatic troglobites did survive beneath the ice sheets. The heat balance between the surface of the ice and the Earth's interior maintained water-filled caves a short distance beneath the sole of the glaciers. Today, for example, we find amphipods in Castleguard Cave, British Columbia, which underlies a present-day glacier.

John Holsinger has studied cave amphipods in the New England states that could not have migrated northward in the time since the glaciers retreated. *Stygobromus borealis,* for example, which lives in Vermont, has no close relatives in the caves farther south. It apparently lived in a water-filled cave refugium beneath the Pleistocene ice sheet until 12,500 years ago, when the ice retreated for the last time from Vermont.

Intergradation with Surface Animals

AFTER THE INITIAL COLONIZATION of a cave by a species adapted to a cavelike surface environment, a separation from the population at the surface is required for evolution of the species in the cave to proceed further. If interbreeding between surface and cave forms occurs freely, there can be no genetic basis for evolution of a new cave-adapted species.

Climatic changes can affect surface animals differently from cave animals, because the geographic range of a species is determined not by the average conditions of its habitat but by the extremes of environmental fluctuation in which it lives. Because environmental fluctuations derived from those of the surface are greatly subdued in caves, the potential geographic range is larger underground for species that tolerate darkness.

A long-term climatic change might increase to intolerable levels the annual incidence of surface freezing or drought, but have relatively little effect in a cave. As a consequence, the range of the surface population of a species could be shifted far enough away from the cave to permit evolution to troglobites to proceed in the cave population.

It is sometimes possible to demonstrate a relationship between troglobitic and surface forms through a series of intermediate types of animals. One of the classic examples of such a transition is that between characid fish found in Chica Cave and those in the nearby Tampaón River in the state of San Luis Potosí, Mexico. *Astyanax mexicanus*, living in the Tampaón, has normal eyes and is normally pigmented. In the dark zone of Chica Cave the most nearly related

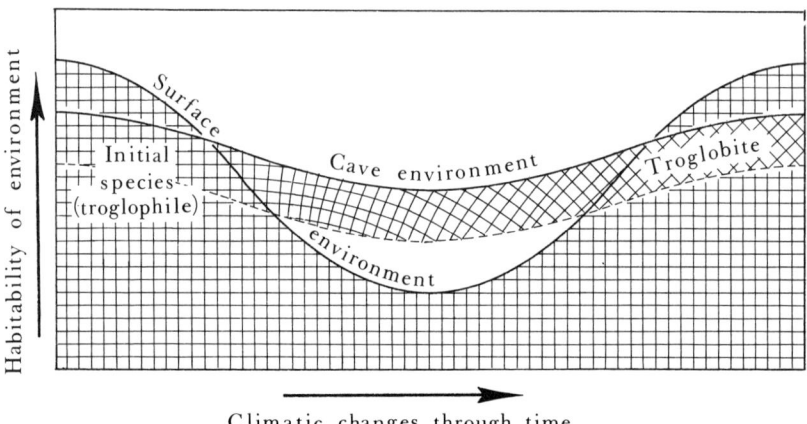

Origin of a cave-restricted species during a long-term climatic change. Because seasonal environmental fluctuation is subdued in a cave, the cave population of an initial species living both in and outside the cave may survive while the surface population dies out locally. By the time climatic conditions again permit descendents of the surface species to repopulate the area of the cave, adaptive genetic changes and the lack of interbreeding have led to a distinct cave species, a troglobite.

species is *Astyanax jordani*, a nonpigmented, blind form. Intermediate types, inhabiting other nearby caves, are not totally blind. Although their optic organs are defective, they can develop an imperfect retinal image, and they demonstrate a strong schooling instinct—traits that are missing in the totally blind *Astyanax jordani*. Fish with even this imperfect vision tend to rest quietly, whereas the blind fish seem to wander aimlessly. In the light, fish of a sighted species often attempt to school with *Astyanax jordani*; these attempts occasionally end with the sighted fish killing the blind ones.

Crosses in captivity between *Astyanax mexicanus* and *A. jordani* result in forms intermediate in both anatomy and behavior patterns. When the intermediate forms themselves are crossed, increased variation results, with extremes ranging from pigmented blind fish to nonpigmented sighted fish. This shows that several genes are involved in the development of these traits.

Reproductive Separation of Cave and Surface Relatives

DURING THE PAST 20 YEARS, Francis Howarth, Stewart Peck, and others have found that blind and colorless cave animals are common in the tropics. This fact seems to call into question the hypothesis, developed in northern regions, that cave species are climatic relicts—that is, that cave populations diverged from their surface ancestors because a change in climate locally killed off those at the surface. Therefore, speleologists have launched a search to find other mechanisms to explain the lack of interbreeding between new cave species and their surface parents.

To begin with, however, while great ice sheets repeatedly waxed and waned nearer the Earth's poles, the tropics were not totally free from climatic change. For example, dusty ice that formed during the Pleistocene is preserved in cores from Andean glaciers and shows that the Amazon Basin was dryer when the northern ice sheets were at their maximum. Nevertheless, the distribution pattern of cave animals in both tropical and temperate regions suggests that climate change is not the only process that can create troglobites.

The boundary between a cave and the surface seems to be a potent barrier across which surface species can spin off cave species that then remain effectively isolated from the gene pool of their parent.

One way in which a parent species might lose this breeding contact is through disruption of the biologic clock of its incipient new daughter species. Biochemical processes control natural cycles in organisms that do not necessarily correspond with daily or yearly cycles. They need to be reset repeatedly by exposure to sunlight and the seasons, just as a human traveler can shorten the discomfort of jet lag by early-morning exposure to the sun. Animals whose biologic clocks diverge from surface cycles because of the darkness and isolation of caves are likely to have different fertile periods than their surface relatives. This can immediately isolate an incipient cave species. Then, over time, genetic drift can equip it with the further characters of a full troglobite.

Cycle Lengths that Distinguish Cave and Surface Isopod Species
(*After* René Ginet.)

Property	Stenasellus (Cave species)	Stenasellus (Surface species)
Female carries embryos	9–10 months	3 weeks
Juvenile period	5–7 years	3 months
Adult period	7 or more years	6–8 months
Juvenile molts	every 6 months	every 15 days
Adult molts	only 1	every 15 days
Time between reproduction	2 years	1 month

Regressive Characteristics

REDUCED PIGMENT AND DEGENERATE EYES of cave-dwelling animals are called *regressive* characteristics. The surface-dwelling ancestors of these animals had sight and pigment, but those traits were not essential to survival in a cave. Other cave-related traits are progressive. Inasmuch as caves in temperate latitudes have a minimal food supply, progressive adaptations can take the form of a reduced food intake and a lowered pace of living. The cave species have lost certain characteristics that are normal in surface forms. On the following pages we suggest how they did so.

Regressive evolution can occur throughout a whole animal family. One of the largest subfamilies of cave beetles is the Bathysciinae, found in many European caves. More than 600 species and subspecies of bathysciines are regressive, their degree of regressiveness being correlated with their degree of progressive adaptation to life in caves.

Mutations Unchecked by Natural Selection

When considering the cause of the regressive evolution of cave animals, one should remember that many unrelated cave animals have two traits in common—blindness and extremely reduced pigment. Cave animals ranging in the scale of specialization from ostracods and isopods, through insects and spiders, to fish and salamanders, are blind and nearly colorless, whereas their surface relatives can see and are normally pigmented. A general process must therefore be operating to degenerate the pigment no longer needed for camouflage and the eyes no longer needed for seeing.

The theory of natural selection can be used to explain the survival of animals with a character, such as camouflage, that is an advantage in their environment. But it is less effective in explaining the disappearance of such a character when it is no longer needed. Yet both pigment and eyes have degenerated in thousands of species during their adjustment to cave life.

The explanation for this degeneration may be that mutations—abrupt, random changes in genes—tend to be more often destructive than constructive. Geneticists have estimated that only 1 mutation in 500 is beneficial. In a surface environment, natural selection would winnow out blind individuals or those without camouflage. In the darkness of the cave, however, where sight and camouflage are both useless, there tends to be an accumulation of mutant animals lacking both functional eyes and pigment. Mutations causing degeneration of the eyes, for example, are fairly common, whereas there is none that can restore the eyes once they have been lost. The mutation pressure that thus develops is like a one-way street; it causes a steady degradation of organs that darkness has made useless.

Natural selection preserves what is useful despite the destructive mutations, but it does not protect what has ceased to be useful.

An organism does not need all of its genes at the beginning of its development. The mutated genes may come into action at various stages in the development of the individual, and the onset of an abnormality may therefore begin at a late stage. This is nicely illustrated by the European cave salamander *Proteus anguinus*. The young, or larval form, has well-developed, functional eyes and conspicuous spots of coloration, but the full-grown *Proteus* is blind and colorless.

During the evolution of most groups of cave animals, the pigment becomes reduced before the eyes degenerate. No direct mutations from pigmented to albino forms have been observed in cave-adapted species; the pigment is lost in stages. Because even in an animal as simple as the fruit fly more than 40 different genes are required to control the eye color alone, it is easy to see how such a step-by-step regression can take place in caves. C. Kosswig has recently demonstrated that a series of multiple factors is necessary for the inheritance of pigment in the isopod *Asellus aquaticus*, found in some Slovenian caves.

It is possible to imagine a surface species that is already adapted to living in a cool moist environment migrating into a cave by chance. Once a group of these animals is underground, they might benefit by the relief from predatory pressure, hardship of living, and competition from related animals. This could give rise to a new and long-lived species, if the animals were able to continue reproducing, and if interbreeding with surface forms ceased.

While the species is establishing itself, however, natural selection will eliminate features that are handicaps in its new home. Mutations will also tend to eliminate features that are useless but whose absence is not necessarily a handicap. The combined result would be that any organs not positively needed for survival will gradually degenerate. Finally, after a long period of time, the species could change so much that it is no longer be able to survive outside the cave, and then it would be a true troglobite.

Skull of a modern human *Homo sapiens sapiens* compared with fossil relatives. The skull capacity of *Homo sapiens* is about 1500 cubic centimeters, that of *Homo erectus* about 1000 cubic centimeters, and that of *Australopithecus africanus* about 450 cubic centimeters.

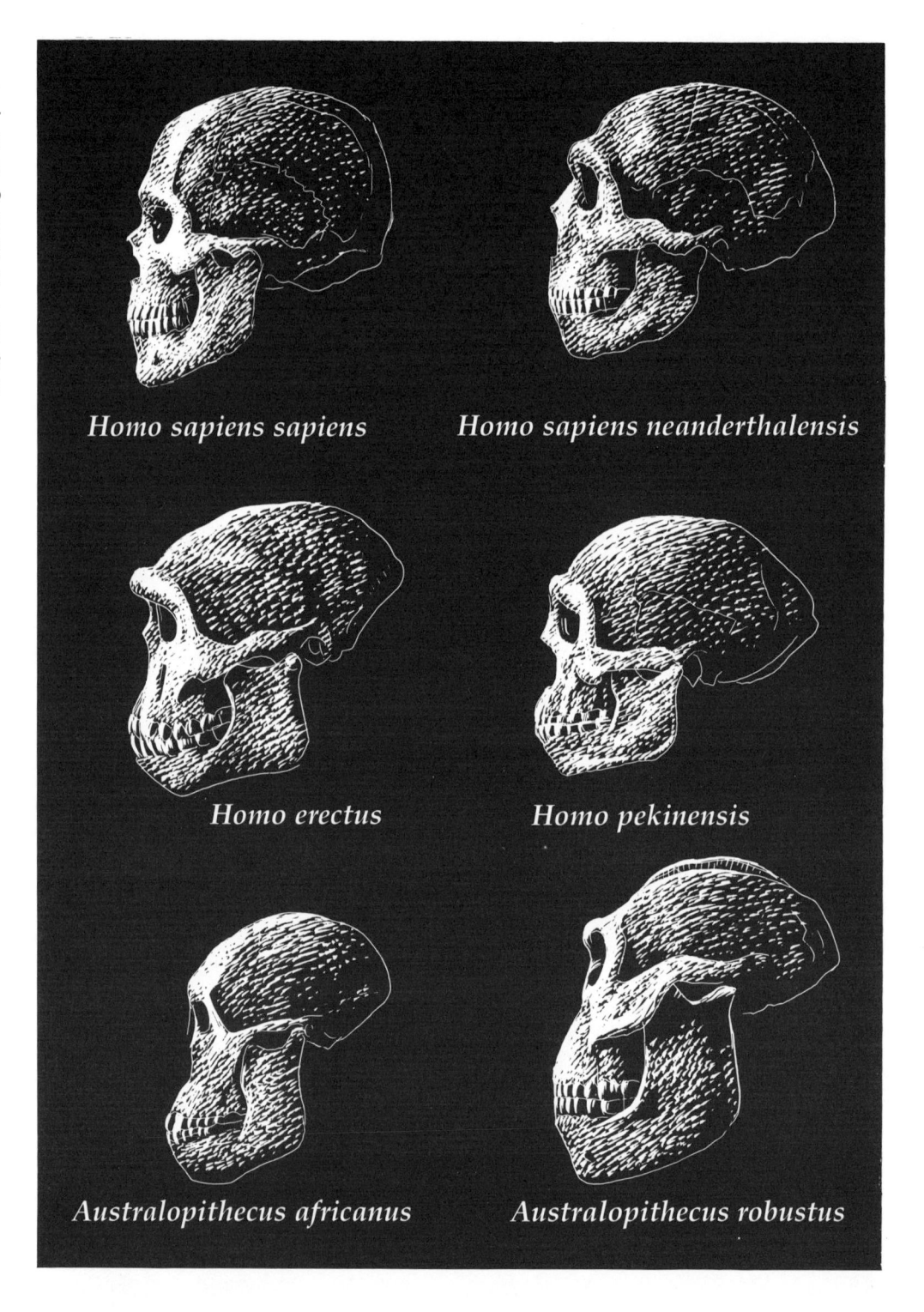

Homo sapiens sapiens

Homo sapiens neanderthalensis

Homo erectus

Homo pekinensis

Australopithecus africanus

Australopithecus robustus

8: Uses of Caves

Human Ancestors

OUR PRESENT KNOWLEDGE of the early development of human beings and their culture is intimately associated with the exploration and study of caves. Many skeletal remains of primitive relatives of *Homo sapiens sapiens* have been found in caves in Africa and Asia, and continuing studies indicate that our subspecies originated in Africa.

An interesting group of early primates are the australopithecines. Bones of *Australopithecus africanus*, a species thought to have become extinct about 2 million years ago, were first discovered in limestone caves broken into by a quarry near Taungs in Botswana. In Sterkfontein Cave, near Johannesburg, South Africa, remains of *Australopithecus africanus* occur in a stratigraphic sequence that also contains simple bone and pebble tools. For many years this was interpreted to mean that the australopithecines used tools, but excavations in Sterkfontein Cave in August 1976 show that the tools are related to younger sediment layers that contain bones of the genus *Homo*.

The australopithecines were 120-150 centimeters tall and weighed about half as much as 170-centimeter-tall modern humans. They apparently walked much as we do and had teeth similar to ours, but their skulls and faces were apelike, and they had only about one third the average brain capacity of a modern human adult.

Remains of a related species *Australopithecus robustus*, with a similar brain size but a larger body, have been found in shelter caves near the Makapan Limeworks, South Africa. *Australopithecus robustus* used these caves as eating places. Thousands of animal bones have been found in association with the bones of this species. Many of the animal leg bones have been shattered, and some investigators think they were cracked open by *Australopithecus robustus* for their marrow.

It is highly improbable that these species of *Australopithecus*, or any other prehuman or human creatures, ever dwelt for long periods deep in caves. Although they may have used the outer parts of caves for shelter and as places in which to eat their meals, they must have found the dark zone of most caves too damp and cool for prolonged habitation. Extensive archeological excavations beyond the twilight zone of numerous caves have failed to disclose major accumulations of occupational debris.

Homo erectus erectus, a fossil found in Java in the late 19th Century, is an intermediate form between *Australopithecus* and modern humans. Remains closely related to *Homo erectus erectus* were excavated in 1930 from a deep cave at Zhoukoudian, a short distance west of Beijing, China. Both this subspecies—*Homo erectus pekinensis*—and *Homo erectus erectus* have brain capacities larger than that of *Australopithecus* and approaching that of *Homo sapiens sapiens*. They made simple tools and presumably were hunters. *Homo erectus pekinensis* is the first subspecies definitely known to have possessed a knowledge of the use of fire.

The steadily enlarging brain and the upright stance of human ancestors were clear advantages, but initially these traits were incompatible. Standing on two legs and swift running require a pelvis that is too small for giving birth to the large skull required by a large brain. Reproductive selection solved this problem by delaying most of the growth of the human head until after birth.

The immature brain of newborn humans, however, makes them far more dependent on long-term parental care than the young of other mammals. This need for extended nurturing led to the adoption of close-knit family and social groups. Ultimately, this closeness was responsible for human culture and society.

The Neanderthals, so called because their remains were first discovered in a cave in the Neander Valley near Düsseldorf, Germany, were definitely human. The remains of this subspecies called *Homo sapiens neanderthalensis* (or archaic *Homo sapiens*) are common in Europe. The Neanderthals occupied shallow caves in which there is little evidence of paintings or other forms of art, although red ocher, identical with that used in the famous paintings made by later races in the caves of southern France, has been found on some bones of Neanderthals from La Chapelle-aux-Saints, France.

Remains of *Homo sapiens neanderthalensis* found in Le Moustier Cave, France, indicate that they were short and stocky with receding chins. Their brain capacities were greater than those of modern humans, and they had more prominent brow ridges. We know nothing about their skin color or degree of hairiness; whatever any drawing or painting of them shows of these traits is based on guesswork.

The Neanderthals were most prosperous during the early part of the last glacial stage of the Pleistocene Epoch. According to the fossil record they disappeared about 30,000 years ago, to be replaced by the Cro-Magnons, who belong, as we do, to the subspecies *Homo sapiens sapiens*. The position of several human races in the cultural sequence of Europe is summarized in the following table, along with the characteristic products of each period, both utilitarian and artistic.

European Cultural Sequences

CULTURE	PREDOMINANT HUMAN RACE	ESTIMATED YEARS AGO	CHARACTERISTIC PRODUCTS
Mousterian	Neanderthal	150,000–30,000	Stone tools
Aurignacian	Cro-Magnon	30,000–20,000	Stone blade and bone tools; earliest cave paintings
Solutrean	Cro-Magnon	20,000–18,000	Pressure-flaked two-faced stone points and knives; stone engravings; wall sculpture
Magdelanian	Cro-Magnon	18,000–10,500	Elaborate stone and bone tools; multicolored paintings; relief figures carved from cave silt

Cave Art

CRO-MAGNON REMAINS, first found in a cave of that name near Les Eyzies, France, have been found at hundreds of sites in continental Europe, as well as in Paviland Cave in South Wales. The Cro-Magnon populations were especially active during the waning phases of the last glacial stage. At that time they were most productive as artists. On rock walls in the dark zone of caves, Cro-Magnons painted thousands of animal figures, many of them extremely realistic in pose and vivid in color. Some of their paint was liquid, and some of it paste. They spread it with their fingers, with brushes made of reeds or hair, or with pads of moss. Sometimes they blew dry pigment onto the wall by means of a hollow tube. Their palettes were large flat bones. Their pigments consisted mainly of red and yellow ocher, probably mixed with animal fat. They obtained pigment from burned bones and from the coatings of black manganese minerals found on many cave walls. For light they used floor fires and pine torches or stone lamps, with marrow or fat for fuel. The lamps had wicks perhaps of moss, and could produce a fairly bright light for several hours.

Some of the earliest cave art consists of finger tracings of geometric designs and simple animal outlines on damp clay-covered walls. Typical of the Aurignacian Period of Paleolithic art, these are the first drawings we know of that were made by human beings for purposes unconnected with their immediate material needs. There was then a gradual development through the Solutrean Period to the Magdalenian, when a high point in prehistoric art was reached. During this period, Cro-Magnon artists made countless bold and realistic multicolored paintings of the animals familiar to them, particularly of those they were in the habit of killing.

On December 18, 1994, speleologists discovered a spectacular new collection of Paleolithic art dated at 30,000 years old in Chauvet Cave, near Vallon-Pont D'Arc, France. It contains animal species never before painted—leopard, hyena, owl—and uses beautiful shading techniques on the bodies of the long-horned rhinoceros and lion. The faces of the lions in particular, males of which are shown to lack manes, are exquisitely drawn, even to such details as their whiskers. This large collection predates the well-known paintings in Lascaux and Altamira Caves by at least 10,000 years. Its art emphasizes dangerous animals such as lions, bears, and rhinos, in contrast with the hunted animals depicted at Lascaux and Altamira.

European cave dwellers also began at approximately this time to make sculptured figures of animals in high relief by carving partially dried cave silt with sharp flint knives. Because of the silt's firm consistency, many of these sculptures have remained intact to the present day. In some caves, footprints and handprints of the early artists are still preserved in the silt. Frequently, also, these sculptors incorporated natural surface irregularities of the cave walls into the contours of their sculptures, and they sometimes represented details by chipping away parts of the rock.

Dated at 30,000 years, this drawing of a long-horned rhinoceros discovered in 1994 in Chauvet Cave, France, is one of the oldest known cave paintings. Its age refutes a former opinion that European cave painting advanced progressively from stick figures to highly refined figures such as this.

In southern Africa, more than 3,000 shelter caves contain paintings and rock engravings. Little is known, however, of the methods of the artists who produced these works, nor has their art been definitely correlated with specific skeletal remains. Giant hand axes and other forms of chipped stone implements have been found at and near many of the caves. Because most of these sites are far from centers of scientific investigation, little study has been devoted to them. When they are studied as intensively as the cave art of Europe has been, they should yield valuable evidence pertaining to our remote relatives.

Explanations for the artistic activity of Paleolithic cave dwellers are varied. Perhaps it was merely a means of spare-time recreation; when game was abundant, even the hunters may have had many hours of the day free for artistic endeavor. Or the artists could have been camp tenders, people too old for hunting, or persons otherwise incapable of more active pursuits.

It is possible that some of the pictures and sculptured figures were made in the hope of ensnaring game by magic, for many of them portray nets, spears, and traps. Other simple symbols may be seasonal markers, and these are commonly grouped with stylized symbols of male and female sex organs. There is a remarkable sameness in the choice of subject, composition, and execution of much of the cave art, even in widely separated caves, indicating some sort of unifying influence. Communication among the early tribes throughout southern Europe probably was extensive.

Examination of Paleolithic art is helpful to paleontologists who are studying Pleistocene vertebrates. It helps them to date the extinction of many animal species. Furthermore, the realistic pictures made by cave artists give information about the form and color of the extinct animals that could not be obtained by merely examining skeletons. It also demonstrates that humans lived at the same time as the species shown in the paintings, and gives an insight into the Cro-Magnons' food habits as well as their methods of obtaining food.

Cave art shows refinements in form, proportion, and rhythm of movement that indicate a high degree of artistic perception. In particular, the intensely dramatic movement illustrated in cave paintings in southern France and northern Spain indicates highly developed imaginative powers. The authenticity of detail that characterizes so much of Paleolithic art also indicates keen observation, a trait that would be extremely valuable to people dependent on hunting for survival.

One of the most remarkable facts about cave art is that it very seldom represents a full human figure. The few that are depicted are so highly stylized that not one of them could ever be identified as the portrait of any particular individual. This lack of portraits is frustrating to anthropologists, who would be very glad to know just what some of the cave dwellers looked like. Perhaps these cave artists were inhibited by a superstition that still prevails in many tribes—the belief that people can be injured or even killed by mutilation of pictures or images that resemble them.

Early Americans

PEOPLE MIGRATED much more recently to the Western Hemisphere than to Europe and Asia, and there is strong evidence that the first inhabitants of America came from Asia. Archeological excavations, especially during the past 40 years, have demonstrated a marked similarity between early cultures in Siberia and Alaska. This is not surprising, given that those areas have been united during much of late geologic time. During the most recent glacial stage, sea level throughout the world was lowered by about 100 meters because of the tying up of an immense volume of water in the ice of continental glaciers. When this happened, the shallow seafloor in and near the Bering Sea was laid bare, so that Asia and North America were connected by a belt of dry land. Although the climate of this belt was cold in the winter, most of it was free from glaciers because the local snowfall was very light. Game was abundant, and the area was as hospitable to human beings as the present subcontinents of Alaska and Siberia, or perhaps even more so.

The movement of people from Asia to North America across this intervening lowland need not be considered a migration. This strip of dry land simply provided an area in which people could live in tolerable comfort, just as adjacent areas have done for the Eskimos in more recent times. Human beings as well as other animals lived on it, and slowly, generation after generation, the area that they occupied in North America became larger. The movement of population probably was not like the westward trek of a frontier family; it resembled, more likely, the gradual encroachment of the suburbs of a large modern city on the surrounding countryside.

We do not yet know with certainty the time of people's first arrival on the North American continent. Many sites in the western United States and northern Mexico provide evidence that people were present and widely distributed 10 to 11 thousand years ago. Whereas archeologists from time to time have reported cultural material believed to date from 20 to as much as 50 thousand years ago, further study invariably has shown that the material either is incorrectly dated or consists of naturally fractured stones that only resemble artifacts. The most compelling evidence that people did not arrive in the Western Hemisphere before 12,000 years ago comes from Europe and other continents that definitely had much longer periods of occupation. At suitable sites in those places, traces left by humans are abundant and unequivocal.

Once the Bering land bridge had emerged from the sea, and suitable passes opened up between glacial ice caps in Canada, humans seem to have dispersed throughout the inviting but previously uninhabited hemisphere within a few centuries.

EXCAVATION IN SANDIA CAVE, in the mountains east of Albuquerque, New Mexico, illustrates the value of modern speleological methods in controlling archeological interpretations. A cultural layer at Sandia Cave contains beautifully fashioned spearheads found under a layer of stalagmitic flowstone dated at 9,000 years ago. The spearheads are the type known as Folsom Points, because they were first found at Folsom, New Mexico. They are amazingly regular in form, and each has a smooth, shallow groove or flute on each face. These were the first New World points to be made by flaking with firm pressure rather than by chipping with sharp blows. The Folsom people who used these fluted points lived from 10,200 to 10,900 years ago, and we know that they hunted the woolly mammoth and now-extinct species of peccary, bison, and camel, because Folsom Points are lodged in the skeletons of some of these animals. A slightly older group, the Clovis people, also first found in New Mexico, made somewhat larger points from 10,900 to 11,200 years ago, and these are the earliest fully documented artifacts in the New World.

The Folsom cultural layer in Sandia Cave overlies a layer of yellow ocher, which takes its color from the mineral goethite, $FeO(OH)$. Microscopic study shows that iron bacteria formed it. A second layer of flowstone lies under the Folsom material and above the ocher.

Over large areas in the entrance passage of Sandia Cave, the ocher layer is missing, and cultural material rests directly on the bedrock floor of the cave. Because the ocher layer is missing near the entrance, investigators in the days before the availability of radiometric dating methods believed that the cultural layer there is older than the ocher and the intervening flowstone and hence much older than the Folsom layer.

In 1936 and later, excavators found several distinctive artifacts in this presumably very old layer that have a barb-like shoulder on one side only. These resemble tools from the Solutrian tradition in Europe dated at 18 to 20 thousand years ago. Hence the archeologists believed that these "Sandia Points" are the oldest in the Americas and that they provide a direct link with Europe.

In 1986, however, Vance Haynes and George Agogino found that deeper inside Sandia Cave, no undisturbed artifacts underlie the ocher layer. New uranium-series dating indicates that the second flowstone layer directly above the ocher is more than 300,000 years old. And careful tracing of the beds back toward the cave entrance shows an abrupt termination of the ocher layer, beyond which lies the presumed Sandia cultural material.

Haynes and Agogino's conclusion is that the Folsom people themselves had mined out the yellow ocher near the cave entrance, probably to use as body paint, either directly or heated to convert it to a vivid red color. Consequently, the Folsom and Sandia cultures are the same. Moreover, the distinctive Sandia Points may have been special tools used in the mining. And the 9,000-year-old younger flowstone, which covered everything and sealed it off after all this activity had ceased, served well to confound the initial archeologists.

Evidence for a connection between North American and Asia has been reinforced by the discovery of points, 12,000 years old, in Ust-Kanskaiya Cave, Siberia, that have chipping similar to that of the Sandia Points. Clovis-like cultural material from the Nenana Valley, Alaska, has been dated at from 11,100 to 11,800 years. Fluted points have been recovered from sites in northwestern Canada, Mexico, Central America, Bolivia, and Ecuador. In Fells Cave, Patagonia, almost at the southernmost tip of South America, points have been found that are characteristic of a culture that existed there 10,000 years ago. And from caves just north of the Straits of Magellan, skeletons dated by the radiocarbon method at 7,000 to 9,000 years old have been excavated.

The use of native plants as well as hunting for food is recorded by the presence of milling stones used to grind burro weed seeds in Danger Cave, Utah. These stones have been suggested by radiocarbon dating of associated carbon-bearing material to be about 10,800 years old. In Bat Cave, New Mexico, corncobs have been recovered that show that corn was cultivated nearby 3,000 to 4,000 years ago.

Some Oregon caves contain woven sandals and fine-twisted basketry that are 9,000 years old. At the entrance to Russell Cave, in northern Alabama, awls, necklaces, rings, and lamps dating back 8,000 years have been collected in such abundance that the total quantity weighs several tons. In the Hueco Mountains of western Texas is a whole series of caves from which ancient mats, baskets, and garments made of woven cloth have been excavated, together with many tools used for cleaning animal skins and for cooking, grinding, and hunting. These finds indicate that an advanced culture existed near these caves about 5,000 years ago.

THE EARLIEST MODERN EXPLORERS of Mammoth Cave, Kentucky, discovered that prehistoric Indians had preceded them into the depths. Lost and discarded artifacts were found for distances of up to 5 kilometers from the natural entrance to the cave. In Salts Cave, whose entrance also is within Mammoth Cave National Park, explorers in 1875 found what has been called an Indian "mummy," a desiccated body dried out in the cave atmosphere. More artifacts and evidence of prehistoric activity came to light there and in Mammoth Cave through the remainder of the 19th and early 20th Centuries.

No pottery has been discovered, but gourds and mussel-shell spoons have been found in the caves. The people also utilized bark fiber, grasses, and other tough stringy plants to weave fabric for articles of clothing. Many moccasins fashioned from this woven material have been found in the caves.

Charred remnants of dried weed stalks and reeds (the same kinds that still grow today on the banks of Green River near Mammoth Cave) make it clear that the Indians used these materials for torches to light their way in the dark. Why did these aborigines travel such long distances into the strange world of these caves? The answer is found on many of the cave walls, which are covered with marks where the Indians chipped off and carried out a thin crust of sulfate minerals.

On June 7, 1935, two cave guides discovered another Indian "mummy" deep inside Mammoth Cave. This is the body of a man about 45 years old and about 160 centimeters tall. He was found approximately 3 kilometers from the natural entrance on a ledge above the passage floor, pinned beneath an enormous breakdown slab. Evidence of nearby chipping indicates that he had been collecting sulfate minerals when the slab fell on him.

In 1957, two radiocarbon determinations were made on prehistoric materials from Mammoth Cave, and the dates obtained—420 B.C. and 280 B.C.—prove that these Indians were working in Mammoth Cave more than 2000 years ago. Beginning in 1963, and continuing to the present, members of the Cave Research Foundation undertook archaeological investigations in Salts Cave and later in Mammoth Cave and other caves of Mammoth Cave National Park. More radiocarbon dates have been obtained, and we now know that the prehistoric miners were exploring and mining for minerals beginning 4,000 years ago and continuing approximately until the time of Christ.

They broke off gypsum, mirabilite, and epsomite from the walls and ceiling wherever they could reach them. The gypsum probably was used to make white body paint. Mirabilite and epsomite may have been desired for food seasoning, as each has a distinctive taste—mirabilite is salty, epsomite bitter. They are also cathartic, so they might have been sought for use as medicine. Still another possibility is that the ancient miners may have thought that these substances, which are produced deep inside the Earth, have magical powers. In any case, the minerals must have been considered valuable, for evidence of barter materials suggests that they were traded to distant tribes, and they were mined during a time span of nearly 2,000 years.

SUPERSTITIONS about caves were formerly prevalent in many parts of the world. For example, in Slovenia, people believed that the white salamanders which occasionally washed out of caves were baby dragons. As recently as the summer of 1936, such superstitions were very real indeed in Romania. The speleologist J.H. Hroch von Dalebor told us that he asked at that time for information about a cave near Rodna, Romania. The villagers readily told him how to reach the cave, but when told that he intended to *enter* it they begged him not to. This region is the birthplace of the legend of the vampires, blood-sucking ghosts in human form, that were believed to dwell in caves and that had long since given their name to the American vampire bat. Disregarding the protests of the villagers, Hroch's group entered the cave and explored it. When they returned to the surface, they found the entrance encircled by solemn-faced men armed with murderous-looking clubs. The village leader then announced that since Hroch and his friends had escaped the vampires in the cave, the speleologists must be vampires themselves. Whereupon the villagers raised their clubs, and the speleologists had to run for their lives.

Modern Uses of Caves

WE DESCRIBE ABOVE some of the uses to which caves have been put during the past. The mining of saltpeter for gunpowder during the War of 1812 was of vital importance to the survival of our young nation. During the Civil War, the Confederates resumed operation of these natural mines when the blockade cut off their foreign sources of gunpowder, and many of the tools they used for this purpose are preserved in the caves today. In Bat Cave, Arizona, extensive mining of bat guano for fertilizer continued until about 1960, and small guano-mining operations continue in other caves today. The product finds its way into novelty shops, where for a price of 1 dollar you can take home a half-pound bag to use on cherished flowers and potted plants.

The list of other uses for caves is long. These uses range from mushroom-growing and cheese-aging to the use of cave air for air-conditioning buildings. The biggest business involving caves is tourism. In the United States alone, over 150 caves are open to the public, and

Saltpeter vats and stirring paddles in Calfkiller Cave, Tennessee. The saltpeter leached from cave silt in these straw-lined vats during the Civil War was used in the production of gunpowder.

several million visitors pass through them each year. Because these caves are skillfully lighted and provided with well-graded trails, they offer by far the easiest way of becoming familiar at first hand with some of the phenomena discussed in this book.

The possible use of limestone caves as shelters from the effects of nuclear weapons received considerable attention for a time. In 1962 a committee of the National Speleological Society completed a study of this possibility. It was concluded that without elaborate alteration, caves would be unsuitable as shelters for large numbers of people. The chief reasons for this conclusion are the following: (1) few caves are near large population centers; (2) water supplies in caves are derived from the surface and are therefore subject to contamination by radioactive fallout; (3) the floor areas of caves are irregular and are commonly covered with large blocks of limestone that have fallen from the ceiling; and (4) some caves "breathe," making it difficult to protect them from contaminated air, while others have poor ventilation, making them dangerous for prolonged occupancy by many people.

We believe that the most important future use of caves will be for research. The extreme simplicity of the cave environment, in which there is no sunlight and temperature is virtually constant, makes a cave a unique laboratory. Combined with these simplifications is the additional advantage that we can often examine the end products of continuous "experiments" that have been running for thousands of years.

Some of the results of past research in the cave-related sciences are summarized on the preceding pages, but who can say what the future holds? Whatever may be in store, countless unexplored caves remain to reveal their beauty to speleologists' lamps and their secrets to ever more refined tools for research.

References and Related Reading

Bakalowicz, M.J., Ford, D.C., Miller, T.E., Palmer, A.N., and Palmer, M.V. 1987. Thermal genesis of dissolution caves in the Black Hills, South Dakota. *Geological Society of America Bulletin* 99: 72-99.

Baker, A., Smart, P.L., Edwards, R.L., and Richards, D.A. 1993, Annual growth banding in a cave stalagmite. *Nature* 364: 518-520.

Baker, A., Smart, P.L, and Edwards, R.L., 1995, Paleoclimate implications of mass spectrometric dating of a British flowstone. *Geology* 23: 309-312.

Balenzano, F., Dell'Anna, L., and Di Perro, M. 1974. Ricerche mineralogiche su alcuni fosfati rinvenuti nelle grotte di Castellana (Bari): strengite, alluminifera, vivanite, taranakite, brushite e idrossiapatite. *Soc. Ital. Min. Petrol.* 30: 543-573.

Balenzano, F., Dell'Anna, L., and Di Perro, M. 1976. Francoanellite, $H_6K_3Al_5(PO_4)_8 \cdot 13H_2O$, a new mineral from the caves of Castellana, Puglia, southern Italy. *Neues Jahrbuch für Mineralogie Monatshefte* 2: 49-57.

Barbour, R.W., and Davis, W.H. 1969. *Bats of America*. Lexington: University of Kentucky Press, 286 p.

Barr, T.C. 1968. Cave ecology and the evolution of troglobites. *Evolutionary Biology* 2: 35-102.

Beck, B.F., ed. 1995. *Karst Geohazards—Engineering and Environmental Problems in Karst Terrane*. Rotterdam: A.A. Balkema.

Beck, B.F., Fram, M., and Carvajal, J.R. 1976. The Aguas Buenas Caves, Puerto Rico: geology, hydrology, and ecology with special reference to the histoplasmosis fungus. *National Speleological Society Bulletin* 38: 1-16.

Becker, H.K. 1925. Zur Kenntnis der Tropfsteine. *Speläologisches. Jahrbuch* 5/6: 117-126.

Bennington, F. 1959. Preliminary identification of crystalline phases in a transparent stalactite. *Science* 129: 1227.

Bernasconi, Reno. 1962. Nota sul giacimento di epsomite di Tana di Val Serrata (Ticino). *Rass. Speleol. Ital.* 14: 362-370.

Blatchley, W.S. 1897. Indiana caves and their fauna. *Indiana Geological Survey Annual Report* 21: 121-212.

Bradbury, J.C. 1959. Crevice lead-zinc deposits of northwestern Illinois. *Illinois Geological Survey Report of Investigations* 210: 1-49.

Bridge, P.J. 1973. Guano minerals from Murra-el-elevyn Cave, Western Australia. *Mineralogical Magazine* 39: 467-469.

Bridge, P.J. 1974. Guanine and uricite, two new organic minerals from Peru and Western Australia. *Mineralogical Magazine* 39: 889-890.

Bridge, P.J. 1975, Urea from Wilgie Mia Cave, W.A. *Western Australian Naturalist* 13(4): 85-86.

Bridge, P.J. 1977. Archerite, $(K,NH_4)H_2PO_4$, a new mineral from Madura, Western Australia. *Mineralogical Magazine* 41: 33-35.

Bridge, P.J., and Clarke, R.M. 1983. Mundrabillaite—A new cave mineral from Western Australia. *Mineralogical Magazine* 47: 80-81.

Bridge, P.J., Pryce, M.W., Clarke, R.M., and Costello, M.B. 1978. Sampleite from Jingemia Cave, Western Australia. *Mineralogical Magazine* 42: 369-371.

Bridge, P.J., and Robinson, B.W. 1983. Niahite—A new mineral from Malaysia. *Mineralogical Magazine* 47: 79-80.

Brigmon, R.L., Martin, H.W., Morris, T.L., Bitton, G., and Zam, S.G. 1994. Biogeochemical ecology of *Thiothrix* ssp. in underwater limestone caves. *Geomicrobiology Journal* 12: 141-159.

Broecker, W.S., Olson, E.A., and Orr, P.C. 1960. Radiocarbon measurements and annual rings in cave formations. *Nature* 185: 93-94.

Brook, G.A., Burney, D.A., and Cowart, J.B. 1990. Desert paleoenvironmental data from cave speleothems with examples from the Chihuahuan, Somali-Chalbi, and Kalahari Deserts. *Palaeogeography, Palaeoclimatology, Palaeoecology* 76: 311-329.

Brucker, R.W., and Watson, R.A. 1987. *The Longest Cave.* Carbondale: Southern Illinois University Press, 316 p.

Cairncross, B., and Dixon, R. 1995. *Minerals of South Africa.* Johannesburg: Geological Society of South Africa. 290 p.

Camacho, A.I., ed. 1992. *The Natural History of Biospeleology.* Madrid: Museo Nacional de Ciencias Naturales. 680 p.

Caumartin, Victor. 1963. Review of microbiology of underground environments. *National Speleolgical Society Bulletin* 25: 1-14.

Chapman, Philip. 1982. The origin of troglobites. *University of Bristol Spelaeological Society Proceedings* 16:133-141.

Chapman, Philip. 1993. *Caves and Cave Life.* London: HarperCollins. 224 p.

Chiesi, M., and Forti, P. 1985. Tre nuovi minerali per le grotte dell'Emilia, Romagna. *Not. Min. Paleontol.* 45: 14-18.

Chirvinsky, P.N. 1925. Tyuyamunite from the Tyuya-Muyun radium mine in Ferghana. *Mineralogical Magazine* 20: 287-295.

Christiansen, Kenneth. 1985. Regressive evolution in Collembola. *National Speleological Society Bulletin* 47: 89-100.

Christiansen, Kenneth. 1992. Biological processes in space and time: Cave life in the light of modern evolutionary theory. In *The Natural History of Biospeleology*. A.I. Camacho, ed. Madrid: Museo Nacional de Ciencias Naturales, p. 453-478.

Claus, Georg. 1962. Data on the ecology of the algae of Peace Cave in Hungary. *Nova Hedwigia* 4: 55-79.

Cody, A.D. 1978. Ruatapu Cave. *New Zealand Speleological Bulletin* 6: 184-187.

Coleman, J.C. 1945. Stalactite growth in the New Cave, Mitchelstown, County Cork. *Irish Naturalists' Journal* 8: 254-255.

Cooper, J.E., and Kuehne, R.A. 1974. *Speoplatyrhinus poulsoni*, a new genus and species of subterranean fish from Alabama. *Copeia* 1974: 486-493.

Culver, D.C. 1976. The evolution of aquatic cave communities. *American Naturalist* 110: 945-957.

Culver, D.C. 1982. *Cave Life: Evolution and Ecology*. Cambridge: Harvard University Press. 189 p.

Culver, D.C., Kane, T.C., and Fong, D.W. 1995. *Adaption and Natural Selection in Caves*. Cambridge: Harvard University Press. 223 p.

Curl, R.L. 1974. Deducing flow velocity in cave conduits from scallops. *National Speleological Society Bulletin* 36: 1-5.

Curl, R.L. 1986. Fractal dimensions and geometries of caves. *Mathematical Geology* 18: 765-783.

Damon, P.H., Sr., ed. 1991. *Caving in America*. Huntsville: National Speleological Society. 445 p.

Davies, W.E. 1958. Caverns of West Virginia: *West Virginia Geological Survey* 19A: 23.

Davies, W.E. 1960. Origin of caves in folded limestone. *National Speleological Society Bulletin* 22: 5-19.

Davies, W.E., and Moore, G.W. 1957. Endillite and hydromagnesite from Carlsbad Caverns. *National Speleological Society Bulletin* 19: 24-25.

Davis, D.G. 1973. Sulfur in Cottonwood Cave, Eddy County, New Mexico. *National Speleological Society Bulletin* 35: 89-95.

Davis, D.G. 1980. Cavern development in the Guadalupe Mountains: a critical review of hypotheses. *National Speleological Society Bulletin* 42: 42-48.

Davis, S.N., and Moore, G.W. 1965. Seimdiurnal movement along a bedrock joint in Wool Hollow Cave, California. *National Speleological Society Bulletin* 27: 133-142.

Davis, W.M. 1930. Origin of limestone caverns. *Geological Society of America Bulletin* 41: 475-628.

Dawkins, W.B. 1874. *Cave Hunting.* London, Macmillan and Company. 455 p.

Diaconu, Gabriel. 1974. Quelques considérations sur la présence de l'anhydrite dans la grotte "Pestera Diana." *Institut Spéologie "Émile Racovitza," Travaux* 13: 191-194.

Dietrich, R.V. 1960. Virginia mineral localities. *Virginia Polytechnic Institute Engineering Experiment Station Bulletin* 138: 1-84.

Dreybrodt, Wolfgang. 1996. Principles of early development of karst conduits under natural and man-made conditions revealed by mathematical analysis of numerical models: *Water Resources Research* 32: 2923-2935.

Ek, C., and Gewelt, M. 1985. Carbon dioxide in cave atmospheres. New results in Belgium and comparison with some other countries. *Earth Surface Processes and Landforms* 10: 173-187.

Elliott, W.R. 1976. New cavernicolous Rhagidiidae from Idaho, Washington, and Utah (Prostigmata: Acari: Arachnida). *Texas Technical University Museum Occasional Paper* 43: 1-15.

Emmons, W.H. 1928. The state and density of solutions depositing metalliferous veins. *American Institute of Mining and Metallurgical Engineers Transactions* 76: 308-320.

Egenmeier, S.J. 1981. Cavern development by thermal waters. *National Speleological Society Bulletin* 43:31-51.

Faust, Burton. 1947. An unusual phenomenon [cave breathing]. *National Speleological Society Bulletin* 9: 52-54.

Feraud, J., Pillard, F., and Vernet, J. 1976. La talmessite $Ca_2Mg(AsO_4)_2 \cdot 2H_2O$ du karst antéalbien a barytine de Luceram (Alpes-Maritimes). *Bull. Soc. Franç. Min. Crist.* 99: 331-333.

Fischbeck, R., and Müller, G. 1971. Monohydrocalcite, nesquehonite, dolomite, aragonite, and calcite in speleothems of the Frankische Schweiz, Western Germany. *Contributions to Mineralogy and Petrology* 33: 87-92.

Ford, D.C., and Williams, P.W. 1989. *Karst Geomorphology and Hydrology.* London: Unwin Hyman, 601 p.

Garavelli, C.L., and Quagliarella, A.F., 1974. Mellite di Grotta Romanelli (Otranto). *Roma Periodico di Mineralogia* 43: 39-49.

Gèze, B., Lagrange, R., and Pobéguin, T. 1956. Sur la nature du revetement occasionnel des parois ou du sol des grottes ("montmilch"). *Paris Academie Science Comptes Rendus* 242:144-145.

Ginet, R., and Decou, V. 1977. *Initiation à la Biologie et à L'écologie Souterraines.* Paris: Delarge, 345 p.

Goldberg, P.S., and Nathan, Y. 1975. The phosphate mineralogy of et-Tabun Cave, Mount Carmel, Israel. *Mineralogical Magazine* 40: 253-258.

Gospodaric, R., and Habic, P. 1976. *Underground Water Tracing.* Ljubjiana: Institute of Karst Research, 309 p.

Graeme, R.W. 1981. Famous mineral localities: Bisbee, Arizona. *Mineralogical Record* 12 (5): 287-291.

Groves, C.G., and Howard, A.D. 1994. Minimum conditions for limestone cave development. *Water Resources Research* 30: 2837-2846.

Grund, A. 1903. Die Karsthydrographie. *Pencks Geographical Abhandlung* 7: 103-200.

Gulden, R.E. 1996. World long/deep cave list as of April 16, 1996. *National Speleological Society News* 54(5, pt. 2): 190-191.

Gurnee, R., and Gurnee, J. 1990. *Gurnee Guide to American Caves.* Closter: R.H. Gurnee, Inc. 288 p.

Halliday, W.R., 1954, Ice caves of the United States. *National Speleological Society Bulletin* 16: 3-28.

Halliday, W.R. 1976. *Depths of the Earth: Caves and Cavers of the United States.* New York: Harper and Row, 432 p.

Harmon, R.S., Thompson, P., Schwarcz, H.P., and Ford, D.C. 1975. Uranium-series dating of speleothems. *National Speleological Society Bulletin* 37: 21-33.

Haynes, C.V., Jr., 1992. Contributions of radiocarbon dating to the geochronology of the peopling of the New World. In *Radiocarbon after four decades*, eds. Taylor, R.E., and others. New York: Springer-Verlag, p. 355-374.

Haynes, C.V., Jr., and Agogino, G.A. 1986. Geochronology of Sandia Cave. *Smithsonian Institution Contributions to Anthropology* 32: 1-32.

Henderson, E.P. 1949. Some unusual formations in Skyline Caverns, Virginia. *National Speleological Society Bulletin* 11: 31-34.

Hess, W.H. 1900. The origin of nitrates in cavern earths. *Journal of Geology* 8: 129-134.

Hill, C.A. 1976. *Cave minerals.* Huntsville: National Speleological Society, 137 p.

Hill, C.A. 1979. Recent anhydrite and bassanite in Big Bend National Park caves. *National Speleological Society Bulletin* 42: 126-127.

Hill, C.A. 1981. Origin of cave saltpeter. *National Speleological Society Bulletin* 43: 110-126.

Hill, C.A. 1987. Geology of Carlsbad Caverns and other caves of the Guadalupe Mountains, New Mexico and Texas. *New Mexico Bureau of Mines and Mineral Resources Bulletin* 117:1-150.

Hill, C.A., and Buecher, R.H. 1992. Nitrocalcite in Kartchner Caverns, Kartchner Caverns State Park, Arizona. *National Speleological Society Bulletin* 54: 14-16.

Hill, C.A., and Ewing, R.C. 1977. Darapskite, $Na_3(NO_3)(SO_4) \cdot H_2O$, a new occurrence in Texas. *Mineralogical Magazine* 41: 548-550.

Hill, C.A., and Forti, P. 1986. *Cave Minerals of the World.* Huntsville: National Speleological Society. 238 p.

Hixson, G.P.J. 1962. A new cave mineral. *National Speleological Society Nittany Grotto Newsletter* 10 (5): 71

Hoch, H., and Howarth, F.G. 1993. Evolutionary dynamics of behavioral divergence among populations of the Hawaiian cave-dwelling planthopper *Oliarus polyphemus* (Homoptera: Fulgoroidea: Cixiidae). *Pacific Science* 47: 303-318.

Holsinger, J.R. 1966. A preliminary study on the effects of organic pollution of Banners Corner Cave, Virginia. *International Journal of Speleology* 2: 75-89.

Holsinger, J.R. 1978. Systematics of the subterranean amphipod genus *Stygobromus* (Crangonyctidae), Part II: Species of the eastern United States. *Smithsonian Contributions to Zoology* 266: 1-144.

Holsinger, J.R. 1988. Troglobites: The evolution of cave dwelling organisms. *American Scientist* 76: 146-153.

Hobbs, H.H., Jr., Hobbs, H.H., III, and Daniel, M.A. 1977, A review of the troglobite decapod crustaceans of the Americas. *Smithsonian Contributions to Zoology* 244: 1-183.

Howarth, F.G. 1973. The cavernicolous fauna of Hawaiian lava tubes. *Pacific Insects* 15: 139-151.

Howarth, F.G. 1987. The evolution of non-relictual tropical troglobites. *International Journal of Speleology* 16:1-16.

Howarth, F.G. 1991. Hawaiian cave faunas: Macroevolution on young islands. In *The Unity of Evolutionary Biology* 1. E.C. Duntley, ed. Portland: Discordies Press, p. 285-295.

Howarth, F.G. 1993. High-stress subterranean habitats and evolutionary change in cave-inhabiting arthropods. *American Naturalist* 142(supplement): 65-77.

Howarth, F.G., and Stone, F.D. 1990. Elevated carbon dioxide levels in Bayliss Cave Australia: Implications for the evolution of obligate cave species. *Pacific Science* 3: 207-218.

Hughes, A.R., and Tobias, P.V. 1977. A fossil skull probably of the genus *Homo* from Sterkfontein, Transvaal. *Nature* 265: 310-312.

Imbrie, J., Mix, A.C., and Martinson, D.G. 1993. Melankovitch theory viewed from Devils Hole. *Nature* 363: 531-533.

Iliffe, T.M. 1986. The zonation model for the evolution of aquatic faunas in anchialine caves. *Stygologia* 2: 2-9.

Jakucs, László. 1977. *Morphogenetics of Karst Regions.* Bristol: Adam Hilger, 284 p.

Jeannel, René. 1943. *Les Fossiles Vivants des Cavernes.* Paris: Gallimard. 321 p.

Jefferson, G.T. 1976. Cave faunas. In *The Science of Speleology*, eds., T.D. Ford and C.H.D. Cullingford. London and New York: Academic Press, p. 359-421.

Jennings, J.N. 1985. *Karst Geomorphology*. Oxford: Basil Blackwell. 293 p.

Johnston, W.D., Jr. 1930. The rate of growth of stalactites. *Science* 72: 298-299.

Jones, C.R. 1965. The limestone caves and cave deposits of Perlis and North Kedah. *Malayan Nature Journal* 19 (1): 21-30.

Kane, T.C., and Poulson, T.L. 1976. Foraging by cave beetles, spatial and temporal heterogeneity of prey. *Ecology* 57: 793-800.

Kane, T.C., and Culver, D.C. 1992. Biological processes in space and time: Analysis of adaptation. In *The Natural History of Biospeleology*. A.I. Camacho, ed. Madrid: Museo Nacional de Ciencias Naturales, p. 377-399.

Kaye, C.A. 1959. Geology of Isla Mona, Puerto Rico, and notes on the age of Mona Passage. *U.S. Geological Survey Professional Paper* 317-C: 141-178.

Kopper, J.S., and Creer, K.M. 1973. Cova dets Alexandres, Majorca: Paleomagnetic dating and archeological inerpretation of its sediments: *Caves and Karst* 15: 13-20.

Kosswig, C. 1963. Genetische Analyse konsruktiver and degenerativer Evolutionsprozesse: *Zoologisch Systematisch Evolutionsforschung Zeitschrift*1:205-239.

Kríz, Martin. 1892. Die Höhlen in den märischen Devonkalken und ihre Vorzeit. *K.K. geol. Reichsanstalt Verh. Jahrbuch* 41: 443-570.

Kuehn, K.A., and Koehn, R.D. 1988. A mycofloral survey of an artesian community within the Edwards Aquifer of central Texas. *Mycologia* 80: 646-652.

Lange, A.L. 1954. Rock temperature distributions underground. *Cave Studies* 6-7: 21-32.

Lange, A.L. 1991. Mapping Jewel Cave—From the surface. *U.S. National Park Service Park Science* 11(2): 6-7.

Lange, A.L. 1993. Karst streams and caverns: Detecting and mapping from the surface. *Geophysics Group Geo-Fontiers* (3): 1-4.

Larsen, E.S. 1921. The microscopic determination of the nonopaque minerals. *U.S. Geological Survey Bulletin* 679: 116.

Lauritzen, S.-E., Haugen, J.E., Løvlie, R., and Gilje-Nielsen. 1994. Geochronological potential of isoleucine epimerization in calcite speleothems. *Quaternary Research*. 41: 52-58.

Lazarev, J.S., and Philenko, G.D. 1976. Geologo-mineralogicheskie osobennosi Gaurdakskoy karstovoy peshchery. *Peshchery* 16: 45-63.

Lovley, D.R., Chapelle, F.H., and Woodward, J.C. 1994. Use of dissolved H$_2$ concentrations to determine distribution of microbially catalyzed redox reactions in anoxic groundwater. *Environmental Science and Technology*.

Løvlie, R., Ellingsen, K.L., and Lauritzen, S.-E. 1995. Paleomagnetic cave stratigraphy of sediments from Hellemofjord, northern Norway. *Geophysical Journal International* 120: 499-515.

Lowry, D.C. 1964. Palygorskite in a cave in New Zealand. *New Zealand Journal of Geology and Geophysics* 7: 917.

Luiszer, F.G. 1994. Speleogenesis of Cave of the Winds, Manitou Springs, Colorado. *Karst Waters Institute Special Publication* 1: 91-109.

Machatschki, V.F. 1929. Original-mitteilungen an die Redaktion. *Centralblatt für Mineralogie* 1929A: 321-332.

Martini, J.E.J. 1978. Sasaite, a new phosphate mineral from West Driefontein Cave, Transvaal, South Africa. *Mineralogical Magazine* 42: 401-404.

Martini, J.E.J. 1980, Mbobomkujlite, hydrombomkulite, and nickelalumite, new minerals from Mbobo Mkulu Cave eastern Transvall. *Geological Survey of South Africa Ann.* 14 (2): 1-10.

Martini, J.E.J. 1983. Lonecreekite, sabieite, and clarite, a new secondary ammonium ferric-iron sulphate from Lone Creek Fall Cave, near Sabie, eastern Transvaal. *Geological Survey of South Africa Ann.* 17:29-34.

Martini, J.E.J. 1984. Additional contributions to the mineralogy of the Transvaal caves. *South African Speleological Assiciation Bulletin* 22: 35-40.

Martini, J.E.J. 1991. Swaknoite [Ca(NH$_4$)$_2$(HPO$_4$)$_2$·H$_2$O, orthorhombic]: A new mineral from Arnhem Cave, Namibia. *South African Speleological Society Bulletin* 32: 72-74.

Martini, J.E.J., and Kavalieris, I. 1978. Mineralogy of the Transvaal caves. *Geological Society of South Africa Transactions* 81: 47-54.

Mason-Williams, Anne. 1967. Further investigations into bacterial and algal populations of caves in South Wales. *International Journal of Speleology* 2: 389-395.

Mawson, D. 1930. The occurrence of potassium nitrate near Goyder Pass, McDonnell Ranges, Central Australia. *Mineralogical Magazine* 22: 231-237.

McClurg, D.R. 1996. *Adventure of Caving*. Carlsbad: D&J Press, 251 p.

McKee, E.D. 1930. Vanadinite in the Grand Canyon. *Grand Canyon Nature Notes* 4: 52.

Merrill, G.P. 1894. On the formation of stalactites and gypsum incrustations in caves. *U.S. National Museum Proceedings* 17: 77-81.

Miotke, F.-D., and Palmer, A.N. 1972. *Genetic Relationship Between Caves and Landforms in the Mammoth Cave National Park Area.* Hannover: Geographischen Institut, 69 p.

Mitchell, R.W., and Reddell, J.R. 1973. Studies on the cavernicole fauna of Mexico and adjacent regions. *Association of Mexican Cave Studies Bulletin* 5: 1-201.

Mohr, C.E., and Poulson, T.L. 1966. *The Life of the Cave.* New York: McGraw-Hill, 232 p.

Moore, G.W. 1952. Hollow hemispherical cave deposits. *National Speleological Society Stanford Grotto Monthly Report* 2:31-35.

Moore, G.W. 1952. Speleothem—a new cave term. *National Speleological Society News* 10(6): 2.

Moore, G.W. 1954. The origin of helictites. *National Speleological Society Occasional Papers* 1: 1-16.

Moore, G.W. 1958. Role of earth tides in the formation of disk-shaped cave deposits. *2nd International Speleological Congress* (Bari) *Proceedings* 1: 500-506.

Moore. G.W. 1960. Geology of Carlsbad Caverns, New Mexico. *National Speleological Society Guide Book* 1: 10-18.

Moore, G.W. 1961. Dolomite speleothems. *National Speleological Society News* 19: 82.

Moore, G.W. 1962. The growth of stalactites. *National Speleological Society Bulletin* 24: 95-106.

Moore, G.W. 1981. Manganese deposition in limestone caves. *8th International Speleological Congress* (Bowling Green) *Proceedings* 2: 642-644.

Moore, G.W. 1994. When will we have an accepted explanation for cave nitrate deposits? *Karst Waters Institute Special Publication* 1: 53-54.

Moore, G.W. 1994. Speleology: Creatures from the black lagoon. *Nature* 369: 100.

Moore, G.W., and Bukry, D. 1968. Electronmicrograph of moonmilk. *National Speleological Society News* 26: 121-122.

Morinaga, H., Yonezawa, T., Adachi, Y., Inokuchi, H., Goto, H., and Yaskawa, K. 1994. The possibility of inferring paleoseismicity from paleomagnetic dating of speleothems, western Japan. *Tectonophysics.* 230: 241-248.

Murray, J.W., and Dietrich, R.V. 1956. Brushite and taranakite from Pig Hole Cave, Giles County, Virginia. *American Mineralogist* 41: 616-626.

Mylroie, J.E., and Carew, J.L. 1995. Geology and karst geomorphology of San Salvador Island, Bahamas. *Carbonates and Evaporites* 10: 193-206.

Northup, D.E., Carr, D.L., Crocker, M.T., Cunningham, K.I., Hawkins, L.K., Leonard, P., and Welbourn, W.C. 1994. Biological investigations in Lechuguilla Cave. *National Speleological Society Bulletin* 56: 54-63.

Ogg, J.G., 1995. Magnetic polarity time scale of the Phanerozoic. In *Global Earth Physics: a Handbook of Physical Constants,* ed. Ahrens, T.J. Washington: American Geophysical Union. p. 240-270.

Padera, K., and Povondra, P. 1964. Das Vorkommen des Huntits und Magnesits aus den Grotten Zbrasov bei Teplice nad Becvou. *Univ. Carolinae Acta Geologica* 1964: 15-24.

Palmer, A.N. 1975. Origin of maze caves. *National Speleological Society Bulletin* 37: 57-76.

Palmer, A.N. 1988. Solutional enlargement of openings in the vicinity of hydraulic structures in karst regions. *Association of Ground Water Scientists and Engineers Proceedings* 2:3-15.

Palmer, A.N. 1991. Origin and morphology of limestone caves. *Geological Society of America Bulletin* 103: 1-21.

Palmer, A.N. 1995. Geochemical models for the origin of macroscopic solution porosity in carbonate rocks. *American Association of Petroleum Geologists Memoir* 63: 77-101.

Papamarinopoulos, S., Readman, P.W., Maniatis, Y., and Simopoulos, A. 1991. Paleomagnetic and mineral magnetic studies of sediment from Ball's Cavern, Schoharie, U.S.A. *Earth and Planetary Science Letters* 102: 198-212.

Parés, J.M., and Pérez-González, A. 1995. Paleomagnetic age for hominid fossils at Atapuerca archaeological site, Spain. *Science* 269: 830-832.

Peck, S.B. 1974. The invertebrate fauna of tropical American caves, Part 2: Puerto Rico, an ecological and zoogeographic analysis. *Biotropica* 6: 14-31.

Peck, S.B. 1986. Evolution of adult morphology and life-history characters in cavernicolous *Ptomophagus* beetles. *Evolution* 40: 1021-1030.

Peck, S.B. 1990. Eyeless arthropods of the Galapagos Islands, Ecuador: Composition and origin of the cryptozoic fauna of a young, tropical oceanic archipelago. *Biotropica* 20: 366-381.

Peck, S.B. 1993. Galapagos Islands troglobites: The questions of tropical troglobites, parapatric distributions with eyed-sister species, and their origin by parapatric speciation. *Mémoires de Biospéologie* 20:19-37.

Pérez Martínez, J.J., and Wiggen, R.W. 1953. Los depósitos de fosforitas de Sabanas Hidalgo y Ayancual, estado de Nuevo León. *México Inst. Nac. Inv. Minerales* 32: 1-33.

Peterson, D.W., Holcomb, R.T., Tilling, R.I., and Christiansen, R.L. 1994. Development of lava tubes in the light of observations at Mauna Ulu, Kilauea Volcano, Hawaii. *Bulletin Volcanologique* 56:343-360.

Plummer, L.N., Wigley, T.M.L., and Parkhurst, D.L. 1978. The kinetics of calcite dissolution in CO_2–water systems at 5° to 60°C and 0.0 to 1.0 atm CO_2. *American Journal of Science* 278: 179-216.

Pohl, E.R. 1955. Vertical shafts in limestone caves. *National Speleological Society Occasional Papers* 2: 1-24.

Polyak, V.J., and Mosch, C.J. 1995. Metatyuyamunite from Spider Cave, Carlsbad Caverns National Park, New Mexico. *National Speleological Society Bulletin* 57: 85-90.

Poulson, T.L. 1972. Bat guano ecosystems. *National Speleological Society Bulletin* 34: 55-59.

Poulson, T.L. 1992. The Mammoth Cave ecosystem. In *The Natural History of Biospeleology*. A.I. Camacho, ed. Madrid: Museo Nacional de Ciencias Naturales, p. 569-611.

Poulson, T.L. 1994. Long-term ecological monitoring at Mammoth Cave. *Cave Research Foundation Newsletter* 22(3): 10-12.

Poulson, T.L., and White, W.B. 1969. The cave environment. *Science* 165: 971-981.

Ransome, F.L. 1904. Geology and ore deposits of Bisbee Quadrangle, Arizona: *U.S. Geological Survey Professional Paper* 21: 1-168.

Reinbacher, W.R. 1994. Is it gnome? Is it berg? Is it mont? Is it mond? An updated view of the origin and etymology of moonmilk. *National Speleological Society Bulletin* 56: 1-13.

Repenning, C.A. 1992. *Allophaiomys* and the age of the Olyor Suite, Krestovka Sections, Yakutia. *U.S. Geological Survey Bulletin* 2037: 98 p.

Repenning, C.A., and Grady, F. 1988. The microtine rodents of the Cheeta Room fauna, Hamilton Cave, West Virginia, and the spontaneous origin of *Synaptomys*. *U.S. Geological Survey Bulletin* 1853: 32 p.

Rickard, T.A. 1924. The Ahumada lead mine and the ore deposits of the Los Lamentos Range, in Mexico. *Engineering and Mining Journal* 118: 365-374.

Roberts, R.J., and Arnold, D.C. 1965. Ore deposits of the Antler Peak Quadrangle, Humboldt and Lander Counties, Nevada. *U.S. Geological Survey Professional Paper* 959-B: 19.

Rogers, B.W. 1975. Selected caves in the Great Basin Ranges. *National Speleological Society Guidebook* 16: 63.

Rogers, B.W. 1976. Melanterite found in Soldiers Cave, California. *National Speleological Society News* 34: 108-109.

Rogers, B.W., and Williams, K.M. 1982. Mineralogy of Lilburn Cave, Kings Canyon National Park, California. *National Speleological Society Bulletin* 44: 23-31.

Roosevelt, A.C., Lima da Costa, M., Llpes Machado, C. Michab, M., Mercier, N., Valladas, H. Feathers, J. Barnett, W., Imazio da Silviera, M., Henderson, A. Sliva, J, Chernoff, B., Reese, D.S., Holman, J.A., Toth, N., and Schick, K. 1996. Paleoindian cave dwellers in the Amazon: The peopling of the Americas. *Science* 272: 373-384.

Sarbu, S.M., and Popa, R. 1992. A unique chemoautotrophically based cave ecosystem. In *The Natural History of Biospeleology*. A.I. Camacho, ed. Madrid: Museo Nacional de Ciencias Naturales, p. 637-666.

Sarbu, S.M., Kane, T.C., and Kinkle, B.K. 1996. A chemoautotophically based cave ecosystem. *Science* 272: 1953-1955.

Schadler, J. 1932. Ardealit, ein neues Mineral, $CaHPO_4 \cdot CaSO_4 + H_2O$. *Zentralbatt für Mineralogie* A: 40-41.

Schmidt, V.A. 1982. Magnetostratigraphy of sediments in Mammoth Cave, Kentucky. *Science* 217: 827-829.

Schwarcz, H.P., Harmon, R.S., Thompson, P., and Ford, D.C. 1976. Stable isotope studies of fluid inclusions in speleothems and their paleoclimatic significance. *Geochimica et Cosmochica Acta* 40: 657-665.

Seeman, Robert. 1981. Systematik der Mineral-Paragenesen in Österreichischen Hohlen. *8th International Speleological Congress* (Bowling Green) *Proceedings* 1: 256-259.

Seeman, Robert. 1984. Neufunde sekundärer Karbonatmineralisationen in Höhlen des Dachsteins und des Untersberges (Nördiche Kalkalpen). *Die Höhle* 35: 253-262.

Shaub, B.M. 1962. Rhodochrosite: The ornamental banded material from Argentina. *Mineral Digest* 4: 46-56.

Shaw, T.R. 1992. *History of Cave Science*. Broadway: Sydney Speleological Society, 338 p.

Shepard, C.U. 1857. Nitrammite. *American Journal of Science* 24: 124.

Simon, R.B. 1973. Cave cricket activity rhythms and the earth tides. *Journal of Interdisciplinary Cycle Research* 4: 31-39.

Sinclair, W.C. 1982. Sinkhole development resulting from ground-water withdrawal in the Tampa area, Florida. *U.S. Geological Survey Water Resources Investigation* 81-50.

Smolianinova, N.N. 1970. Nekotorye dannye po mineralogii i genezisu mestorozhdenia Tyuya-Muyun. Moscow: *Izd. Nauk SSSR Nauka*, 56-90.

Spencer, L.J. 1908. Ore hopeite and other zinc phosphates and associated minerals from the Broken Hill Mines, North-western Rhodesia. *Mineralogical Magazine* 15: 1-38.

Stringham, Bronson. 1946. Tinticite, a new mineral from Utah. *American Mineralogist* 31: 395-400.
Sullivan, Nicholas. 1960. Checklist of macroscopic troglobitic organisms of the United States. *American Midland Naturalist* 64: 123-160.
Sullivan, Nicholas. 1962. Observations on the population dynamics of a cavernicolous ecosystem. *University of Notre Dame PhD Dissertation*, 150 p.
Sullivan, Nicholas. 1967. Ecology of the Río Camuy Cave area, Puerto Rico. *National Speleological Society Bulletin* 29: 38-39.
Sullivan, Nicholas. 1969. Observations on the behavior and longevity of cavernicolous Rhaphidophoridae (Orthoptera). *5th International Speleological Congress* (Stuttgart) *Proceedings* 5: 1-7
Sullivan, Nicholas. 1974. Biology and ecology of the El Convento cave-spring system (Puerto Rico). *International Journal of Speleology* 6: 109-114.
Sullivan, Nicholas. 1982. Australia—The 1982 Chillagoe expedition. *Speleologist* 1982: 231-232.
Sullivan, Nicholas. 1988. Chillagoe cave fauna—challenges. *17th Australian Speleological Federation Conference* (Tropicon, Cairns). 94-97.
Sullivan, Nicholas. 1995. Scientific exploration of Chillagoe Caves. *Karst Waters Institute Conduit 3(2): 3.*
Sweeting, M.M. 1972. *Karst Landforms.* London: Macmillan Press, 362 p.
Sweeting, M.M., and Pfeffer, K.H., eds. 1976. Karst processes. *Zeitschrift Geomorphologie Supplement* 29: 1-210.
Szabo, B.J., Kolesar, P.T., Riggs, A.C., Winograd, I.J., and Ludwig, K.R. 1994. Paleoclimatic inferences from a 120,000-yr calcite record of water-table fluctuation in Browns Room of Devils Hole, Nevada. *Quaternary Research* 41: 59-69.
Sztrókay, K.I. 1959. Mineralogische Beobachtungen aus der Aggteleker Tropfsteinhöhle (Ungarn). *Die Höhle* 10: 50-56.
Taylor, W.J. 1858. Mineralogical notes: Lecontite. *Philadelphia Academy of Natural Sciences Proceedings* 10: 172-174.
Thrailkill, John. 1971. Carbonate deposition in Carlsbad Caverns. *Journal of Geology* 79: 683-695.
Thrailkill, John. 1976. Speleothems. *Developments in Sedimentology* 20: 73-86.
Thrailkill, J., and Robl, T.L. 1981. Carbonate geochemistry of vadose water recharging limestone aquifers. *Journal of Hydrology* 54: 195-209.
Tower, G.W., and Smith, G.O. 1899. Geology and mining industry of the Tintic District, Utah. *U.S. Geological Survey Annual Report* 19: 697.

Tsykin, R.A., and Tsykina, Z.L. 1979. Polezyne iskopaemye peshcher vostochnoy chasti al'taesayanskoy gornoy obl'asti. Perm: *Ispil'zovanie peshcher-Tezisy dokl'adov seminara*, p. 65-66.

Valladas, H., Cachier, H., Maurice, P., Bernaldo de Quiros, F., Clottes, J., Cabrera Valdés, Uzquiano, P., and Arnold, M. 1992. Direct radiocarbon dates for prehistoric paintings at the Altamira, El Castillo, and Niaux Caves. *Nature* 357: 68-70.

Vandel, A. 1965. *Biospeleology*. Oxford: Pergamon Press, 524 p.

Walker, T.L. 1918. Mineralogy of the Hudson Bay Mine, Salmo, British Columbia. *University of Toronto Studies in Geology* 10: 1-25.

Walker, T.L. 1919. Stalactitic barite from Madoc, Ontario. *American Mineralogist* 4 (7): 79-80.

Wang, X. 1982. Phosphate minerals in the karst cavities adjacent to Guilin, Guangxi Provence. *Acta Mineralogica Sinica* 2: 153-156.

Warwick, G.T. 1950. Calcite bubbles—a new cave formation. *National Speleological Society Bulletin* 12: 38-42.

Watson, P.J. 1969. The prehistory of Salts Cave, Kentucky. *Illinois State Museum Reports of Investigations* 16: 1-86.

Watson, P.J., ed. 1974. *Archeology of the Mammoth Cave area*. New York: Academic Press, 272 p.

Watson, P.J. 1986. Cave archaeology in the Eastern Woodlands. *Anthropology of the Americas Masterkey* 59(4):19-25.

Watson, R.A., and Smith, P.M. 1971. Underground wilderness. *International Journal of Environmental Studies* 2: 217-220.

Watson, R.A., and White, W.B. 1985. The history of American theories of cave origin. *Geological Society of America Centennial Special Volume* 1: 109-123.

Weller, J.M. 1927. The geology of Edmonson County. *Kentucky Geological Survey* 28: 70.

White, W.B. 1958. The occurrence of celestite in Cumberland Caverns. *National Speleological Society News* 16: 57.

White, W.B. 1971. Sulfate minerals in central Kentucky karst. *Cave Research Foundation Annual Report* 13: 12.

White, W.B. 1988. *Geomorphology and Hydrology of Karst Terrains*. Oxford: Oxford University press. 464 p.

White, W.B., and Deike, G.H., 1962. Secondary mineralization in Wind Cave, South Dakota. *National Speleological Society Bulletin* 24: 74-87.

White, W.B., and White, E.L., eds. 1989. *Karst Hydrology: Concepts from the Mammoth Cave area*. New York: Van Nostrand Reinhold. 346 p.

Wilkens, Horst. 1992. Neutral mutations and evolutionary progress. In *The Natural History of Biospeleology*. A.I. Camacho, ed. Madrid: Museo Nacional de Ciencias Naturales, p. 401-422.

Caves in the United States Open to the Public

Alabama
De Soto Caverns, Childersburg
Rickwood Caverns, Warrior
Russell Cave National Monument, Bridgeport
Sequoyah Caverns, Valley Head

Arizona
Colossal Cave, Vail
Grand Canyon Caverns, Dinosaur City

Arkansas
Blanchard Springs Caverns, Mountain View
Buffalo National River Caves, Harrison
Civil War Cave, Bentonville
Cosmic Cavern, Berryville
Crystal Dome Caverns, Dogpatch
Devil's Den Cave, West Fork (sandstone cave)
Hurricane River Cave, Pindall
Mystic Caverns, Dogpatch
Onyx Cave, Eureka Springs
War Eagle Cavern, Rogers
Wonderland Cave, Bella Vista

California
Boyden Cave, Kings Canyon National Park
Crystal Cave, Sequoia National Park
California Caverns, Mountain Ranch
Lake Shasta Caverns, O'Brien
Lava Caves, Lava Beds National Monument (lava tubes)

Mercer Caverns, Murphys
Mitchell Caverns, Essex
Moaning Cave, Vallecito
Pinnacles Caves, Pinnacles National Monument (talus caves)
Subway Cave, Manzanita Lake (lava tube)
Sunny Jim Cave, La Jolla (sea cave)

Colorado
Cave of the Winds, Manitou Springs

Florida
Florida Caverns, Marianna

Georgia
Rolater Park Cave, Cave Spring

Hawaii
Thurston Lava Tube, Volcano
Waianapanapa Cave, Koeleku (lava tube)

Idaho
Idaho's Mammoth Cave, Shoshone (lava tube)
Lava Caves, Craters of the Moon Mational Monument (lava tubes)
Minnetonka Cave, St. Charles
Shoshone Indian Ice Cave, Gooding (lava tube)

Illinois
Cave-in-Rock Cave, Cave-in-Rock
Upton Cave, Savanna

Indiana
Bluespring Caverns, Bedford
Cave River Caves, Campbellsburg
Donaldson Cave, Mitchell
Marengo Cave, Marengo
Porter Cave, Paragon
Squire Boone Caverns, Corydon
Wolf Cave, Spencer
Wyandotte Cave, Wyandotte

Iowa
Crystal Lake Cave, Dubuque
Horse Thief Cave, Anamosa
Maquoketa Caves, Maquoketa
Spook Cave, McGregor

Kentucky
Carter Caves, Olive Hill
Crystal Onyx Cave, Cave City
Diamond Caverns, Park City
Jesse James Cave, Park City
Mammoth Cave, Mammoth Cave National Park
Mammoth Onyx Cave, Horse Cave

Maine
Anemone Cave, Acadia National Park (sea cave)

Maryland
Crystal Grottoes, Boonsboro

Michigan
Bear Cave, Buchanan

Minnesota
Mystery Cave, Spring Valley
Niagara Cave, Harmony

Missouri
Bluff Dwellers Cave, Noel
Bridal Cave, Camdenton
Cameon Cave, Hannibal
Cathedral Cave, Leasburg
Cave Springs Onyx Cavern, Van Buren
Crystal Cave, Springfield
Crystal Caverns, Cassville
Fantastic Caverns, Springfield
Fantasy World Caverns, Eldon
Fisher Cave, Sullivan
Honey Branch Cave, Elkhead

Indian Burial Cave, Osage Beach
Jacobs Cave, Versailles
Mark Twain Cave, Hannibal
Marvel Cave, Branson
Meramec Caverns, Stanton
Onondaga Cave, Leasburg
Onyx Mountain Caverns, Newburg
Ozark Caverns, Camdenton
Ozark Wonder Cave, Noel
Rock Bridge Cave, Columbia
Round Spring Cave, Round Spring
Talking Rocks Cavern, Kimberly City
Truitts Cave, Lanagan

Montana
Lewis and Clark Cavern, Whitehall
Pictograph Caves, Billings (sandstone caves)

Nebraska
Robbers Cave, Lincoln (sandstone cave)

Nevada
Lehman Caves, Great Basin National Park

New Hampshire
Lost River, North Woodstock (talus caves)
Polar Caves, Plymouth (talus caves)

New Mexico
Carlsbad Cavern, Carlsbad Caverns National Park
Ice Cave, Grants (lava tube)
Slaughter Canyon Cave, Carlsbad Caverns National Park

New York
Howe Caverns, Howes Cave
Ice Caves, Cragsmoor (talus caves)
Natural Bridge Caverns, Natural Bridge
Natural Stone Bridge and Caves, Pottersville
Secret Caverns, Cobleskill

North Carolina
Black Rock Cliffs Cave, Linville
Boones Cave, Troutman (granite cave)

Ohio
Indian Trail Caverns, Carey
Ohio Caverns, West Liberty
Olentangy Indian Caverns, Delaware
Perry Cave, Put-in-Bay
Seneca Caverns, Bellevue
Seven Caves, Bainbridge
Zane Caverns, Bellefontaine

Oklahoma
Alabaster Caverns, Freedom (gypsum cave)

Oregon
Lava River Cave, Bend (lava tube)
Oregon Caves, Oregon Caves National Monument
Sea Lion Caves, Florence (sea cave)

Pennslyvania
Coral Caverns, Manns Choice
Crystal Cave, Kutztown
Indian Caverns, Spruce Creek
Indian Echo Caverns, Hummelstown
Laurel Caverns, Uniontown
Lincoln Caverns, Huntingdon
Lost River Caverns, Hellertown
Onyx Cave, Hamburg
Penns Cave, Centre Hall
Woodward Cave, Woodward

Puerto Rico
Cueva Clara, Parque de las Cavernas del Rio Camuy
La Cueva de Camuy, Camuy

South Dakota
Bethlehem Cave, Bethlehem
Black Hills Caverns, Rapid City
Crystal Cave, Rapid City
Jewel Cave, Jewel Cave National Monument
Rushmore Cave, Keystone
Sitting Bull Cave, Rapid City
Stagebarn Cave, Piedmont
Wind Cave, Wind Cave National Park
Wonderland Cave, Tilford

Tennessee
Bell Witch Cave, Adams
Bristol Caverns, Bristol
Cheroke Caverns, Solway
Cumberland Caverns, McMinnville
Forbidden Caverns, Sevierville
Indian Cave, Blaine
Jewel Cave, Dickson
Lostsea Cave, Sweetwater
Motlow Cave, Lynchburg
Raccoon Mountain Caverns, Chattanooga
Ruby Falls Cave, Chattanooga
Ruskin Cave, Dickson
Tuckaleechee Caverns, Townsend
Wonder Cave, Monteagle

Texas
Cascade Caverns, Boerne
Cave Without A Name, Boerne
Caverns of Sonora, Sonora
Inner Space Cavern, Georgetown
Longhorn Cavern, Burnet
Natural Bridge Caverns, San Antonio
Wonder World Cave, San Marcos

Utah
Timpanogos Cave, Timpanogos Cave National Monument

Virginia
Battlefield Crystal Caverns, Strasburg
Caverns at Natural Bridge, Natural Bridge
Cudjos Caverns, Cumberland Gap (Tenn.)
Dixie Caverns, Salem
Endless Caverns, New Market
Grand Caverns, Grottoes
Luray Caverns, Luray
Massanutten Caverns, Keezletown
Shenandoah Caverns, New Market
Skyline Caverns, Front Royal

Washington
Ape Cave, Amboy (lava tube)
Gardner Cave, Metaline Falls

West Virginia
Lost World Caverns, Lewisburg
Organ Cave, Ronceverte
Seneca Caverns, Riverton
Smoke Hole Cavern, Moorefield

Wisconsin
Cave of the Mounds, Blue Mounds
Crystal Cave, Spring Valley
Eagle Cave, Muscoda
Kickapoo Indian Caverns, Wauzeka

The Authors

GEORGE W. MOORE, who received his doctorate from Yale University in 1960, had a long career as a research geologist with the United States Geological Survey. In 1987, he joined the Department of Geosciences in Oregon State University. He is noted for having coined the word *Speleothem* and is coauthor of the benchmark 1960 Conservation Policy of the National Speleological Society. He contributed to the development of the theory of plate tectonics and has published more than 170 papers in geology and speleology. Dr. Moore is a former president of the National Speleological Society.

NICHOLAS SULLIVAN, F.S.C., is a biospeleologist who has studied cave biota in 1,600 caves in over 100 countries. He is a member of the Christian Brothers teaching institute *Fratres Scholarum Christianarum*. Brother Nicholas received his doctorate from the University of Notre Dame in 1962. His efforts since 1981 have concentrated on the caves of Chillagoe, a rather remote area of northern Queensland, where he has led twenty expeditions. To date, over 25 new taxa of troglobites have been described from Chillagoe. He has also participated in Explorers Club expeditions to the Alta Verapáz of Guatemala and the Rio Camuy System of western Puerto Rico. Dr. Sullivan is a past president both of the National Speleological Society and of the Explorers Club.

The Artist

JOHN SCHOENHERR is a graduate of the Pratt Institute and has been a free-lance magazine and book illustrator for many years. He has also been an active and enthusiastic speleologist.

Index

Actinomycetes, 81, 85, 89
Aerated zone, 15
Aggressive. *See* Undersaturated
Agogino, G.A., 142
Air currents, 38
Air Jernih Cave, Malaysia, 3
Alabama,
    Russell Cave, 143
    Tumbling Rock Cave, 49
Alexandres Cave, Spain, 22
Algae, 81, 121
Alta Verapáz, Guatamala, 170
Altamira Caves, Spain, 138
Amberat, 83
Ammonia, 71
Amphibians, 96
Amphipod, 81, 96, 117, 121, 127
Anchialine cave, 121
Anthodite. *See* Aragonite helictite
Aragonite,
    fossil, 72
    helictite, 61, 62
    origin, 71
    thermometer, 72
Argentina, Fells Cave, 143
Arizona,
    Bat Cave, 145
    Silent River Cave, 64
Aurignacian Culture, 137
Australia, Chillagoe, 170
Australopithecine, 135
Austria,
    Cosanostraloch, 3
    Hirlatz Cave, 3
    Lamprechtsofen Shaft, 3
    Raucherkar Cave, 3
Autotroph, 79
Aven. *See* Vertical shaft

Aven Armand, France, 8

Bacteria,
    anaerobic, 81
    autotrophic, 79
    heterotrophic, 79
    in cave food web, 117
    iron, 81, 142
    manganese, 84
    nitrogen, 82
    oil-field, 21
    sulfur, 81, 120
Bacterial mat, 121
Baker, J.K., 113
Banners Corner Cave, Va., 118
Baradla Cave, Hungary, 81
Barr, T.C., xii, 99
Bat, 113
    banding, 111
    birth, 112
    food, 111
    guano, 70, 111
    in cave food web, 117
    mortality when disturbed, xi
    navigation, 111
    parasites, 95
    vampire, 110
Bat Cave, Ariz., 145
Bat Cave, N.Mex., 143
Bear, 94
Beetle, 92, 99, 100, 121
Béke Cave, Hungary, 81
Belaying, x
Berger Shaft, France, 3
Big Spring, Calif., 42, 43
Biologic clock, 116, 129
Bird, 113
Bird-nest soup, 113

Birnessite, 83, 84
Black Chasm Cave, Calif., 15
Blowing cave, 38
Boj-Bulok Cave, Uzbekistan, 3
Botryoid. *See* Cave coral
Botswana, Taungs Caves, 135
Breathing Cave, Va., 32, 39, 49
Breathing caves, 39
Bristletail, 117
Broecker, W.S., 23

Calcite, 70
    cave bubble, 66
    cave raft, 66
    crystals, 67
    moonmilk, 85
    solubility, 9
    stalactite, 48
Calcium bicarbonate, 9, 48
Calfkiller Cave, Tenn., 146
California,
    Big Spring, 42, 43
    Black Chasm Cave, 15
    Lilburn Cave, 41, 42, 86, 87
    Moaning Cave, 23
    Soldiers Cave, 50, 60, 63
    Windeler Caverns, ii
Calkins, F.C., xii
Cambrian Period, 8
Canopy, 56
Carbohydrate, 121
Carbon dioxide, 15, 16
    from surface soil, 9
    in cave origin, 9
    in rainwater, 16
    in springwater, 16
    in stalactite growth, 48
Carbon-14, origin, 22

Index

Carbonate-hydroxylapatite, 70, 71
Carbonic acid, 9, 15, 49
Carlsbad Cavern, N.Mex., 121
    aragonite, 62, 72
    bat flight time, 113
    cave cricket, 100
    cave pearl, 57
    huntite, 87
    hydrogen sulfide, 20
    hydromagnesite, 87
    limestone age, 8
    moonmilk, 87
    origin, 20
    size, 7
    sulfur, 20
Carpenter Cave, Va., 58
Casteret, Norbert, 58
Catfish, 122
Cathedral Cave, Ky., 102
Caumartin, Victor, 80
Cave,
    accidents, x
    animals, 93
    exploration, x
    food chain, 79, 117
Cave art, 23, 137, 139
Cave blister, 64
Cave bubble, 66
Cave coral, 62, 63
Cave cotton, 64
Cave Creek Caverns, Colo., 35
Cave cricket, 100, 101
    antennae, 102
    eggs, 100
    eye facets, 103
    in cave food web, 117
    tactile organs, 102
Cave exploration, x, 135
Cave fish, 106
    evolution, 128
    food, 105
    range, 126
    reproduction, 105
    saltwater, 122
Cave flower, 64, 65
Cave food web, 117, 118
Cave of the Clouds, Colo., 35
Cave of the Winds, Colo., 35, 62, 84
Cave pearl, 57, 58
Cave raft, 66

Cave rat,
    droppings, 109
    in cave food web, 117
    musk, 83
    urine droplets, 83
    vibrassae, 109
Cave refugium, 127
Cave Research Foundation, 144
Cave silt,
    in cave food web, 117
    origin, 18
    pigment, 81
Cave temperature,
    control, 33
    map, 33
Cave-mineral catalogue, 73
Cavernicole. *See* Cave animal
Caverns of Sonora, Tex., 8, 88, 89
Caves,
    as natural laboratories, 2
    open to the public, 163
    total volume, xi
    United States map, xiv
Ceki Cave, Slovenia, 3
Cenote. *See* Karst pool
Cenote Verde, Mexico, 120, 121
Centipede, 96, 98, 117, 121
Chauvet Cave, France, 138, 139
Chemoautotroph, 79, 121
Cheve System, Mexico, 3
Chica Cave, Mexico, 128
Chillagoe, Australia, 170
Chimney effect, 38
China, Zhoukoudian Cave, 27, 136
Circadian rhythm, 116
Claus, George, 81
Clay mineral, 81
Closed ecosystem, 119
Cockpit karst, 29
Cold-trap caves, 37
Coleman, J.C., 53
Collembolan, 98
Colonization of caves, 125
Colorado,
    Cave Creek Caverns, 35
    Cave of the Clouds, 35
    Cave of the Winds, 35, 62, 84
    Fairy Cave, 35
    Fly Cave, 34, 35
    Flycatcher Cave, 35

    Fulford Cave, 35
    Fulton Cave, 35
    Orient Cave, 35
    Ouray Spring, 84
    Porcupine Cave, 27
    Shoshone Spring, 84
    Spanish Cave, 8, 34, 35
Column, 46, 55
Commercial caves of the United States, 163
Conservation of caves, xi
Contamination, microbial, 80
Copepod, 96, 117, 121
Cosanostraloch, Austria, 3
Cottonwood Cave, N.Mex., 20
Coumo d'Hyouernedo System, France, 3
Crane fly, 98
Crayfish, 96, 116
Creer, K.M., 23
Cretaceous Period, 8
Cricket. *See* Cave cricket
Cro-Magnons, 137
Croatia, Lukina Cave, 3
Crustaceans, 96
Cumberland Bone Cave, Md., 53
Cumberland Caverns, Tenn., 65
Cutter. *See* Rounded fissure
Cyclic animal behavour, 116
Czech Republic, Slouper Cave, 53

Danger Cave, Utah, 143
Dark zone, 115
Darwin, Charles, 4
Davies, W.E., 10, 20
Davis, D.G., 20, 35
Davis, W.M., 10
Dawkins, W.B., 53
Deepest cave in the world, 36
Devonian Period, 8
Diaconu, C., 71
Dissolution grooves, 28
Dogtooth spar, 54
Dolina. *See* Sinkhole
Dolomite, 85
Domepit, 19
Drapery, 55

Earth tides,
    cause of joints, 13
    origin, 116
    sensing by cave animals, 116

Earthworm, 98
Easegill System, United Kingdom, 3
Ebb and flow spring,
    diagram, 44
    mechanism, 41
Ebb and Flow Spring, Mo., 45
Echo location,
    bat, 111
    bird, 114
Edwards Limestone aquifer, Tex., 107
Eldons Cave, Mass., 8
Emmons, W.H., 59
Endellite, 20
Entrance zone, 115
Epsomite, 144
Evolution of troglobites,
    based on climate change, 128
    based on differing biologic clocks, 129

Fairy Cave, Colo., 35
Faust, Burton, 39
Fells Cave, Patagonia, 143
Fish, 96, 106
    evolution, 128
    food, 105
    in cave food web, 117
    range, 126
    reproduction, 105
    saltwater, 122
Fisher Ridge System, Ky., 3
Flatworm, 95, 96, 117, 121
Florida Caverns, Fla., 8
Flowstone, 56
Fly, 98, 117
Fly Cave, Colo., 34, 35
Flycatcher Cave, Colo., 35
Fox, 112
France,
    Aven Armand, 8
    Berger Shaft, 3
    cave art, 138
    Chauvet Cave, 138, 139
    Coumo d'Hyouernedo System, 3
    Jean Bernard System, 3, 36
    Lascaux Cave, 138
    Le Moustier Cave, 136
    Mirolda/Lucien Bouclier Shaft, 3
    Pierre Saint Martin System, 3
Friars Hole, W.Va., 3
Frog, 102

Fulford Cave, Colo., 35
Fulton Cave, Colo., 35
Fungal mat, 121
Fungi, 79, 109, 117, 118, 120

Georgia Republic,
    Sniezhnaja-Mezhonnogo System, 3
    Vjacheslava Pantjukhina Shaft, 3
Ginet, René, 130
Glacière. *See* Ice cave
Glowworm, 98
Gnat, 98, 117
Goethite, 81, 142
Goshute Cave, Nev., 66
Gour. *See* Rimstone dam
Gran Dolina, Spain, 28
Grand Caverns, Va., 53, 62
Grike. *See* Rounded fissure
Grund, Alfred, 10
Guacharo bird, 114
Guano, 28, 70, 71, 98, 99, 111, 145
Guatamala, Alta Verapáz, 170
Guide arrows, xi
Gypsum, 21
    cave blister, 64
    cave cotton, 64
    cave flower, 64
    cave rope, 64
    crystals, 64
    mining by Indians, 144
Gypsum cave,
    Alabaster Caverns, Okla., 167
    Optimisticeskheskaya Cave, Ukraine, 3
    Ozernaya Cave, Ukraine, 3

Hamilton Cave, W.Va., 10, 27
Harvestman, 103, 104
Hawaii,
    dissolution caves, xiv
    lava caves, 127
Haynes, C.V., 142
Heizer, R.F., xii
Helictite, 59, 60, 61
Helmholtz resonator, 40
Hess, W.H., 82
Hetereotroph, 79
Hill, C.A., 20, 82
Hirlatz Cave, Austria, 3
Histoplasmosis, 90

Hölloch, Switzerland, 3
Holsinger, J.R., 118, 127
Holthuis, L.B., 121
Hopeite, 70
Howarth, F.G., 118, 127, 129
Howe Caverns, N.Y., 8
Hroch von Dalebor, J.H., 145
Huautla System, Spain, 3
Hum. *See* Karst tower
Human ancestors, 134, 135
Hungary,
    Baradla Cave, 81
    Béke Cave, 81
Huntite, 85
Hydrogen sulfide,
    in cave ecology, 119
    in cave origin, 20
    in karst pool, 122
Hydromagnesite, 85, 87
Hydrozincite, 70
Hypogene. *See* Hydrogen sulfide

Ice age, 36, 125, 127
Ice cave, 36
Ichneumon fly, 99
Ingleborough Cave, England, 53
Insect, 96
Interbreeding between cave and surface animals, 128, 129
Interior valley, 29
Iotopic dating, 22
Iowa, Weber Cave, 83
Ireland, New Cave, 53
Iron,
    bacteria, 84, 85
    minerals, 84
Isopod, 96, 121, 127
    in cave food web, 117
    life cycle, 130
    marine troglobite, 121
    on sewage, 118
Isotopic dating, 2, 68
Italy, Paolo Roversi Shaft, 3

Jasper Cave, S.Dak., 83
Java fossil hominid, 136
Jean Bernard System, France, 3, 36
Jewel Cave, S.Dak., 3
Johnston W.D., 53
Joints, 12, 62
Jurassic Period, 8

Karren. *See* Dissolution grooves
Karst, 28
  pool, 121
  tower, 29
Kentucky,
  Cathedral Cave, 102
  Fisher Ridge System, 3
  "Great Kentucky Desert," 19
  Mammoth Cave, 3, 7, 8, 24, 25, 100, 143
  Salts Cave, 143, 144
Kopper, J.S., 23
Kosswig, C., 132
Kríz, Martin, 53
Kunsky, Josef, 62
Kurz, Peter, xii

La Cueva de Camuy, P.R., 78
Laminako Aterneko Leizea, Spain, 3
Lamprechtsofen Shaft, Austria, 3
Lange, A.L., 36
Lapaiz. *See* Dissolution grooves
Lascaux Cave, France, 138
Lava cave, 31, 37, 127
Le Moustier Cave, France, 136
Leach, 121
Lechuguilla Cave, N.Mex., 3, 20, 21, 121
Lehman Caves, Nev., 18, 62
  cross section, 14
  limestone age, 8
Lewis, W.C., xii
Lice, 95
Lilburn Cave, Calif., 41, 42, 86, 87
Limestone, 7
Limonite. *See* Goethite
Lublinite, 88
Luiszer, F.G., 84
Lukina Cave, Croatia, 3
Luray Caverns, Va., 8

Magdelanian Culture, 137
Magnesite, 85
Makapan Limeworks, South Africa, 135
Malaysia, Air Jernih Cave, 3
Mammoth Cave, Ky., 3
  archeology, 143
  cave cricket, 100
  cross section, 19
  limesone age, 8
  paleomagnetism, 24, 25
  size, 7

Manganese,
  bacteria, 83, 84, 85
  minerals, 83
Marble, 7
Maryland, Cumberland Bone Cave, 53
Massachusetts, Eldons Cave, 8
Massanutten Caverns, Va., 64
Matts Cave, W.Va., 84
Meadow mouse fossils, 26, 27
Medesan, A., 71
Merrill, G.P., 59
Methane, 21
Metric-English equivalents, xiii
Mexico,
  Cenote Verde, 120, 121
  Cheve System, 3
  Chica Cave, 128
  Naica Cave, 68
  Purificación System, 3
  Tampaón River, 128
Microorganisms,
  harmful, 90
  hydrogen sulfide food, 21
  in food web, 91
  in moonmilk, 85
Midge, 122
Millipede, 96, 97, 98, 117, 121
Miotke, F.-D., 26
Mirabilite, 144
Mirolda/Lucien Bouclier Shaft, France, 3
Mischungskorrosion. *See* Mixture dissolution
Mississippian Period, 8
Missouri, Ebb and Flow Spring, 45
Mite, 103
Mixture dissolution, 17
Moaning Cave, Calif., 23
Model Cave, Nev., 36
Mold, 80
Moldova, Zolushika Cave, 3
Mondmilch. *See* Moonmilk
Mondmilchloch, Switzerland, 88
Monetite, 70
Montmorillonite, 81
Moonmilk,
  color, 88
  crystal structure, 88
  medical use, 90
  microbial constituents, 85
  mineralogy, 85
  origin of tiny grains, 88

  scanning electron micrograph, 86, 87
  transmission electron micrograph, 89
Moore, E.J., xii
Moore, G.W., 20, 170
Morris, T.L., 120, 121
Mosquito, 94, 98, 117
Mosquito fish, 122
Moth, 94, 99
Mousterian Culture, 137
Movile Cave, Romania, 22, 119
Muierii Cave, Romania, 71
"Mummy" in cave, 143
Muscovite, 81
Mutations unchecked by natural selection, 131

Naica Cave, Mexico, 68
National Speleological Society, ii, 147
Neanderthals, 28, 136
Nesquehonite, 85
Nevada,
  Goshute Cave, 66
  Lehman Caves, 8, 14, 62
  Model Cave, 36
  Northumberland Cave, 61
  Whipple Cave, 46
New Cave, Ireland, 53
New Mexico,
  Bat Cave, 143
  Carlsbad Cavern,
    7, 8, 20, 57, 62, 72, 87, 100, 113, 121
  Cottonwood Cave, 20
  Lechuguilla Cave, 3, 20, 21, 121
  Sam (San Antonio Mountain) Cave, 27
  Sandia Cave, 84, 141
New York, Howe Caverns, 8
New Zealand, Waitomo Cave, 98
Nicholas, B.G. *See* Sullivan, Nicholas
Niter, 82
Nitrocalcite, 82
Nitrogen minerals, 71
Northumberland Cave, Nev., 61

Obligative cavernicole. *See* Troglobite
Ohio, Perry Cave, 8
Oil-field bacteria, 21
Ojo Guarena System, Spain, 3
Oomycetes, 120
Opilionids, 103
Optimisticeskheskaya Cave, Ukraine, 3
Ordovician Period, 8
Oregon Caves, Ore., 8

Orient Cave, Colo., 35
Ostracod, 96
Oulopholite. *See* Cave flower
Ouray Spring, Colo., 84
Owl, 26
Oxygen isotope analysis, 69
Ozernaya Cave, Ukraine, 3

Paleolithic art, 139
Paleomagnetic dating, 22, 23, 27, 28
Palmer, A.N., 18, 26
Paolo Roversi Shaft, Italy, 3
Parasites, 95
Parés, Josep, 28
Partings, 12, 62
Paviland Cave, Wales, 137
Peck, S.B., 129
Peking fossil hominid, 136
Pennsylvanian Period, 8
Pérez-Gonsález, Alfredo, 28
Permian Period, 8
Perry Cave, Ohio, 8
Peterson, D.W., 31
Petroleum, in cave origin, 21
pH, 20, 49
Phantom midge, 122
Phosphorus minerals, 71
Phreatic zone. *See* Saturated zone
Pierre Saint Martin System, France, 3
Planarian, 95
Pleistocene Epoch, 26, 125, 137
Pohl, E.R., 20
Polya. *See* Interior valley
Ponor. *See* Sinking stream
Porcupine Cave, Colo., 27
Postojna Cave, Slovenia, 52
Poulson, T.L., xii, 19, 100
Pozo del Madejuno, Spain, 3
Precambrian, 8
Protozoan, 79, 95, 117, 121
Pseudoscorpion, 117, 121
Psychology of explorers, 5
Puerto Rico,
 La Cueva de Camuy, 78
 Rio Camuy System, 170
Purificación System, Mexico, 3
Pyrite, 81

Quaternary Period, 8

Raccoon, 94, 103, 112

Radiocarbon dating, 22, 23, 143, 144
Rappelling, x
Rattlesnake, 109
Raucherkar Cave, Austria, 3
Regressive evolution, 130
Reinbacher, W.R., 88
Relative humidity, 37
Repenning, C.A., 26, 27
Reverse-chimney effect, 38
Rimstone dam, 58
Ringtail, 112
Rio Camuy System, P.R., 170
Rock-climbing technique, x
Romanechite, 83
Romania,
 Movile Cave, 22, 119
 Muierii Cave, 71
Rotifer, 95, 121
Rounded fissures, 28
Russell Cave, Ala., 143
Russia, Ust-Kanskaiya Cave, 143

Salamander, 102, 107, 108, 124
 delay of regressive characters, 131
 in cave food web, 117
 larval development, 107
 sensitivity to light, 127
Saltpeter, 82, 145, 146
Salts Cave, Ky., 143, 144
Sam (San Antonio Mountain) Cave, N.Mex., 27
Sandia Cave, N.Mex., 84, 141
Sandstone cave, 30
Sarbu, S.M., 119
Saturated zone, 14
Scallops, 6, 11, 12
Schmidt, V.A., xii, 24, 25
Schoenherr, J.C., 170
Sea cave, 30
Segregation of animals, 114
Shield, 62
Shoshone Spring, Colo., 84
Siebenhengste Hohgant System, Switzerland, 3
Silent River Cave, Ariz., 64
Silurian Period, 8
Simon, R.B., 102, 103
Sinkhole, 28
 collapse, 29
 debris, 117, 118
Sinking stream, 16

Sitting Bull Cave, S.Dak., 67, 68
Skunk, 94, 103, 112
Skyline Caverns, Va., 62
Slouper Cave, Czech Republic, 53
Slovenia,
 Ceki Cave, 3
 karst, 28
 Postojna Cave, 52
Slug, 95
Snail, 95, 96, 121
Snake, 109, 112
Sniezhnaja-Mezhonnogo System, Georgia Republic, 3
Soil, carbon dioxide from, 9
Soldiers Cave, Calif., 50, 60, 63
Solutrean Culture, 137
South Africa,
 Makapan Limeworks, 135
 Sterkfontein Cave, 135
South Dakota,
 Jasper Cave, 83
 Jewel Cave, 3
 Sitting Bull Cave, 67, 68
 Wind Cave, 3
Space biology, 4
Spain,
 Alexandres Cave, 22
 Altamira Caves, 138
 cave art, 138
 Del Trave System, 3
 Gran Dolina, 28
 Huautla System, 3
 Laminako Aterneko Leizea, 3
 Ojo Guarena System, 3
 Pozo del Madejuno, 3
 Tito Bustillo Cave, 23
 Torco dos los Rebecos, 3
Spanish Cave, Colo., 8, 34, 35
Speleothem,
 definition, 47
 record of ancient climate, 68
Sphalerite, 70
Spider, 96, 102, 103, 117, 121
Spores, 117
Springtail, 98, 103, 117, 121
Stalactite, 49, 52
 breakage, 50
 crystal structure, 50, 53, 54
 growth rate, 52
 spherical, 63
 strength, 50
 tubular, 50

Stalagmite, 55
   crystal structure, 56
   flat-topped, 56
   growth rate, 23
   isotopic dating, 69
   radiocarbon dating, 22
Sterkfontein Cave, South Africa, 135
Stratification of animals, 115
Stream alluvium, 117
Sulfur, 21, 121
Sulfuric acid, 20, 21, 121
Sullivan, Nicholas, 170
Supersaturation, 17
Superstitions about caves, 145
Swallet. *See* Sinking stream
Switzerland,
   Hölloch, 3
   Mondmilchloch 88
   Siebenhengste Hohgant System, 3

Table Mountain Cave, Wyo., 51
Tampaón River, Mexico, 128
Taungs Caves, Botswana, 135
Tennessee,
   Calfkiller Cave, 146
   Cumberland Caverns, 65
Tertiary Period, 8
Texas,
   Caverns of Sonora, 8, 88, 89
   Edwards Limestone aquifer, 107
Thorium-230 dating, 68
Ticks, 95, 103
Tito Bustillo Cave, Spain, 23
Torca dos los Rebecos, Spain, 3
Tower karst, 29
Trave System, Spain, 3
Triassic Period, 8
Troglobite, 94
   from the sea, 121
   hydrogen sulfide, 119

   origin, 128, 132
   relation to ice age, 127
   species in North Ameica, 96
   tropical, 129
Troglophile, 94
Trogloxene, 94
Trout Cave, W.Va., 27
Tubular stalactite, 50, 51
Tumbling Rock Cave, Ala., 49
Turbulent water flow, 13
Twilight zone, 115

Ukraine,
   Optimisticeskheskaya Cave, 3
   Ozernaya Cave, 3
Ulfeldt, S.R., 42, 43
Undersaturation, 14, 17, 49
United Kingdom,
   Easegill System, 3
   Ingleborough Cave, 53
   Paviland Cave, 137
United States,
   cave-temperature map, 34
   known limestone caves, xiv
Uranium isotope dating, 68
Ust-Kanskaiya Cave, Siberia, 143
Utah, Danger Cave, 143
Uzbekistan, Boj-Bulok Cave, 3

Vadose zone. *See* Aerated zone
Vampire bat, 110
Vampire ghost, 145
Vertebrate cave animals, 105
Vertical shaft, 19
Virginia,
   Banners Corner Cave, 118
   Breathing Cave, 32, 39, 49
   Carpenter Cave, 58
   Grand Caverns, 53, 62
   Luray Caverns, 8

   Massanutten Caverns, 64
   Skyline Caverns, 62
Vjacheslava Pantjukhina Shaft, Georgia Republic, 3
Voles, fossil, 26
Volume of world's caves, xi

Waitomo Cave, New Zealand, 98
Wallrock minerals, 117
Warwick, G.T., 66
Wasp, 99
Water scorpion, 121
Water table, 11, 14, 17
Watson, P.J., xii
Watson, R.A., xii
Weber Cave, Iowa, 83
West Virginia,
   Friars Hole, 3
   Hamilton Cave, 10, 27
   Matts Cave, 84
   Trout Cave, 27
Whipple Cave, Nev., 46
White, W.B., xii, 13
Wilson, W.L., 120, 121
Wind Cave, S.Dak., 3
Windeler Caverns, Calif., ii
Wyoming, Table Mountain Cave, 51

X-ray diffraction analysis, 47, 83

Zhoukoudian Cave, China,
   fossil hominids, 136
   fossil voles, 27
Zolushika Cave, Moldova, 3
Zonation of animals, 115